D0205786

Receiving Systems Design

Receiving Systems Design

Stephen J. Erst

Copyright© 1984. Second Printing, April, 1985.
ARTECH HOUSE, INC.
610 Washington St., Dedham, MA

International Standard Book Number: 0-89006-135-1
Library of Congress Catalog Card Number: 84-070222

CONTENTS

PREFACE

This book is the result of many comments which have expressed a desire for a text on receiving systems design. Most of the readers have been exposed to the basics involved but have never put it all together. This text is intended to lead the reader through typical cases from which variations can be made to suit a particular need. For those who may desire to refresh themselves in the basics, a review is presented for reference.

The organization of this book is arranged to address the objective of receiving systems design, with supporting explanations in the following chapters.

INTRODUCTION

This book is intended to assist the reader in the design of receiving systems of four fundamental types:

Down converter
Up converter
Hybrid up and down converter
Wadley up converter

The text consists of five parts presented in the following sequence:

A basic overview of signal characteristics (Chapter 1)
The superheterodyne (Chapter 4)
Components (Chapter 5)
Specialized receiving systems (Chapter 6)
Design examples (Chapter 7)

Interspersed throughout are computer programs written in the BASIC language, to assist the designer in system performance prediction.

The designer should accumulate a library of available components and their characteristics for ready reference. Generally it is most expeditious to procure components rather than undergo design and development efforts of these items, unless the designer has this capability available. This is recommended for initial modeling, later moving to in-house designs if cost effective.

A final chapter includes examples and the sequence of computations and considerations leading to the final design. It is almost always a necessity to revise the structure, as unforseen design faults are found through subsequent performance analysis.

Experience will provide the designer with an insight into what can be done. Low noise and high third order intercept performance, almost always specified, are not simultaneously achievable. A design is usually a compromise of these characteristics.

1

AN OVERVIEW OF SIGNAL CHARACTERISTICS

This section is concerned with the reception of the signal from a distant emitter. Considered are the prediction of the signal strength and the attenuation due to free space path loss. The Fresnel zones are defined for link calculation and the subject of fade margin is addressed. With these basic considerations the link performance can be predicted.

1.1 RECEIVER INPUT POWER PREDICTIONS

To determine the necessary receiver noise figure and sensitivity if it has not been previously specified, it becomes necessary to estimate the signal strength at the receiving antenna. Having determined this, the receiver and antenna requirements can be determined. While this is readily done for line of sight links, it becomes less defined for ionospheric reflection, troposcatter, knife edge diffraction systems, *et cetera*, and will not be discussed here. Most modern links are line of sight limited because of operation at UHF, VHF, and microwave frequencies, which penetrate the ionosphere and are not, therefore, reflected back to earth as HF signals are.

To make this calculation the signal strength at the receiving antenna is

$$P_r = P_t + A_t - \text{path loss} \qquad (1\text{-}1)$$

where

   P_r is the power received at the receiving antenna
   P_t is the transmitter power
   A_t is the transmitting antenna gain in the receiving direction
Path loss is discussed in section (1.2).

The $P_t + A_t$ term is the effective radiated power in the direction of the receiving antenna.

Receiver sensitivity or $P_r(min)$ will have been determined from considerations of S/N, C/N, E_b/N_o, *et cetera*, attainable noise figure, and the receiving antenna gain requirements A_r.

A_r becomes

$$A_r = P_{r\,(min)} - P_r \; (dB \; notation) \tag{1-2}$$

Example: Find the required receiving antenna gain when given:

Path loss = 170 dB

P_t = 100 watts, 50 dBm

A_t = 10 dB

$P_{r(min)}$ = -100 dBm

then the effective radiated power is:

$ERP = P_t + A_t$

$$= 50 + 10 = 60 \; dBm \tag{1-3}$$

and

$P_r = ERP - Path \; loss$

$= 60 \; dBm - 170 \; dB = -110 \; dBm$

$$A_r = -100 \; dBm - (-110 \; dBm) = 10 \; dBm \tag{1-4}$$

The receiving antenna gain must be 10 dB, minimum.

These equations can be manipulated to determine any one parameter knowing the others.

For a discussion of path loss, see section 1.2.

1.2 FREE SPACE PATH LOSS

Electromagnetic emission from a point source radiates energy equally in all directions. At any distance d away from the source, this energy is distributed evenly over a spherical area whose radius is d and its center is the source. It follows that if the transmitted power is P_t, the power per unit area at a distance d is:

$$\frac{P_t}{4\pi d^2} \tag{1-5}$$

The power received by a receiver with an antenna whose effective area is A is:

$$P_r = \frac{P_t}{4\pi d^2} \, A \tag{1-6}$$

Since isotropic antennas are the reference standard upon which antennas are usually compared, it is convenient to utilize this as the receiving antenna. The effective area of the isotropic antenna is

$$A = \frac{\lambda^2}{4\pi} \tag{1-7}$$

where

λ is the wavelength, $\lambda = \dfrac{velocity\ of\ propagation}{frequency}$

Substituting into (1-6) we have

$$P_r = \frac{P_t\,(\lambda^2/4\pi)}{4\pi d^2} \tag{1-8}$$

$$= \frac{P_t\,\lambda^2}{4^2\,\pi^2\,d^2} = \frac{P_t\,\lambda^2}{157.9\,d^2}$$

The path loss in dB is

$$L_p = 10\,\log\,\frac{P_t}{P_r} = 10\,\log\left(\frac{157.9\,d^2}{\lambda^2}\right) \tag{1-9}$$

If d is in miles, and λ is in centimeters, then equation (1-9) becomes

$$L_p = 10\,\log\left(\frac{157.9\,d^2}{\lambda^2}\ \frac{1}{(6.214\cdot 10^{-6})^2}\right)$$

$$= 10\,\log\left(4.089\cdot 10^{12}\ \frac{d^2}{\lambda^2}\right) \tag{1-10}$$

In dB notation,

$$L_p = 126.12\ \text{dB} + 20\,\log d - 20\,\log \lambda \tag{1-11}$$

where

d is in miles

λ is in centimeters

Other forms of this basic equation may be obtained by substituting frequency in GHz for λ.

Then

$$L_p = 92.45 + 20\,\log f + 20\,\log d \tag{1-12}$$

where

d is in kilometers

f is frequency, in GHz

and

$$L_p = 96.58 + 20\,\log f + 20\,\log d \tag{1-13}$$

where

d is in statute miles

f is in GHz

Note that all of the path loss equations assume isotropic receiving antennas. Where the transmitting or receiving antenna has gain, this must be accounted for as a reduction of path loss.

Equation (1-13) is shown in graphical form in Fig. (1-1), for reference purpose.

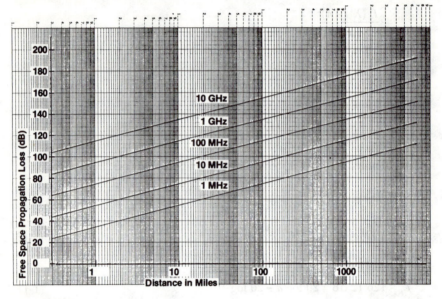

Fig. 1-1. Free space propagation loss

The reader is cautioned in the use of these equations for link calculations. These are free space equations with no intervening obstructions or signal reflections, resulting in multipath situations. If by the use of elevated antennas with moderate gain a free space situation can be approached, then these equations are valid.

For frequencies greater than 8 GHz the environmental effects on the signal must be accounted for. Reference [1] treats this subject in detail.

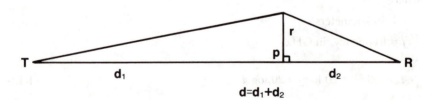

Fig. 1-2. Physical relationship between transmitter T and receiver R where r is the radius of the first Fresnel zone.

1.3 FRESNEL ZONES

Fresnel zones describe the phase behavior of a signal originating at a transmitter T and appearing at a distant receiving site R. With the aid of Fig. (1-2), T and R are connected by a line $n\lambda$ long describing the shortest distance between them. A plane perpendicular to this line is constructed at p and a circle is drawn on this plane containing all points where the path length has increased by $(1/2)$ λ to $(n+1/2)\lambda$. This is the first Fresnel zone which contains nearly 25% of the signal power within its boundary. (This is the most important zone.) Similarly, other circles may be drawn for path length increases of multiples of $(1/2)$ λ. These are successively known as the second, third, fourth, *et cetera*, zones for path length increases of λ, 1.5λ, and 2λ, respectively. All odd multiples of $(1/2)\lambda$ are in phase at R, while even multiples, which are in phase with each other, are out of phase with the odd multiples at R. The signal contributions of each zone are nearly equal, diminishing very slightly as the zone numbers become large. A successive summation of the signal contributions of each zone (i.e., 1,1+2, 1+2+3, 1+2+3....n) would show a cyclic behavior until, with a sufficiently large n, the cyclic amplitude diminishes and the signal at R becomes equal to the free space value.

The first zone is the most important zone and it should be kept clear of obstructions. The radius of this zone at any point along the axis may be found from

$$r = 13.16 \left(\lambda \ \frac{d_1 d_2}{d} \right)^{\frac{1}{2}} \tag{1-14}$$

where

    r is the radius of the first zone (feet)
    λ is the wavelength (cm)
    d_1 is the distance to point p from the transmitter (miles)
    d_2 is the distance to point p from the receiver (miles)
    d is the straight line distance between transmitter and receiver (miles)

In other units, where d is in miles and F is in MHz:

$$r = 2280 \, (d_1 d_2 / dF)^{\frac{1}{2}} \tag{1-15}$$

The value of r maximizes when point p is midway between T and R, at which time r may be found from:

$$r = 1140 \, (d/F)^{\frac{1}{2}} \tag{1-16}$$

Reflection from the earth will vary in magnitude as a function of the reflection coefficient. Where the angle of incidence is small, this coefficient approaches unity. The incidence and reflection angles are equal. There is a phase reversal at the point of reflection for all polarizations. The resulting signal intensity profile for various clearances is shown in Table 1-1. Shown are the cases of reflection from highly reflective, relatively smooth ground and water, and are labeled plane earth and smooth sphere diffraction. The knife edge diffraction case is applicable to fairly smooth vegetated terrain without atmospheric disturbances. In plane earth theory, 6 dB signal enhancement is possible at clearances equal to odd integral multiples of the Fresnel radius.

Table 1-1.

Radio Wave Propagation as Affected by Path Clearance (dB) [2]

Clearance First Fresnel Zone Radius	Knife Edge Diffraction	Smooth Sphere	Plane Earth
-3	-26	>-70	>-70
-2.5	-24	>-70	>-70
-2	-22	-70	>-70
-1.5	-19	-59	>-70
-1	-17	-45	>-70
- .5	-12	-12	>-70
0	0±1	-30	-70
.5		0	0
1.0		+6	+6
1.5		*	*
2.0		*	*
2.5		*	*

1.4 FADE MARGIN

A link is subject to degradation of the signal because of physical changes in the transmission medium, geometry, or both. An allowance for such changes must be made to guarantee the communications reliability of the link. This allowance is established in dB and is called the fade margin. Link reliability is generally expressed in percent values such as 99.9%, which allow an outage of 0.1%.

Multipath fading is a major cause of outages, and is particularly severe in mobile installations. The mechanism is one of reflection or attenuation of the signal from: buildings, water, trees, *et cetera*. The summation of all signals arriving via different paths at the receiving antenna causes enhancement or reduction of the signal phases. These variations are largely random and are called Rayleigh fading because of their distribution. Fig. (1-3) relates link reliability to relative signal strength and is a theoretical maximum. Using this graph, a system with 99% link reliability would require a design signal strength 18 dB above threshold.

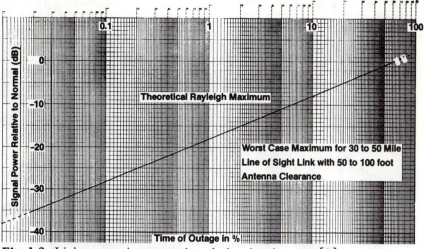

Fig. 1-3. Link outage time *versus* the relative signal power [2].

Example: For a given signal to noise ratio the minimum signal strength is –97 dBm. To meet a link reliability of 99.9%, the system must have a 28 dB fade margin. This gives the receiver a signal strength of –97+28 = –69 dBm.

Because of the shorter wavelengths, higher frequencies are more prone to multipath fading which approaches the limit of Fig. (1-3), above 4 GHz. The effect of frequency on Rayleigh fading is shown in Fig. (1-4) and is a percentage of the maximum shown in Fig (1-3).

Example: For an outage of 0.1% of the time, or a link reliability of 99.9%, a frequency of 1 GHz will have a fade depth of 67% of 28 dB or 18.76 dB. Moving the frequency to 4 GHz results in a fade depth of 90% of 28 dB or 25.2 dB, or 6.44 dB more.

Other variables are weather related, such as: temperature inversion, diffraction, scattering or absorption due to moisture, rain, or snow, and temperature itself.

One solution to fading problems is diversity reception. This is based on redundancy and may involve two or more receivers whose outputs are combined. The redundancy may involve the use of receiving antennas at different locations, feeding several receivers tuned to the same signal. The same information may be transmitted on several different frequencies, each of which is received by a receiver or a combination of both. Antenna polarization may be utilized, as well as time, for the system variables.

Needless to say, diversity is an expensive proposition, and the link configuration would require careful economic analysis before a decision regarding its use could be made.

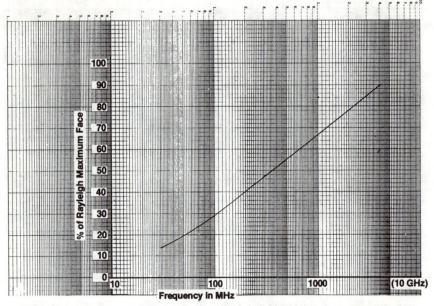

Fig. 1-4. Percent of Raliegh fade maximum *versus* frequency [2].

REFERENCES

[1] Skolnik, Merrill I., *Introduction to Radar Systems*. New York: McGraw-Hill Book Company, 1962.

[2] Bullington, K., "Radio Wave Propagation Fundamentals," *B.S.T.J.*, vol. 36, no. 3, Fig. 8.

2

MODULATION

Key to the design of a receiving system is a knowledge of the characteristics of the signal to be received. The nature of the modulation determines the type of demodulator which must be employed and whether or not signal limiting is required. The frequency range of the modulation defines the post-detection frequency response required. The magnitude of the modulation, together with the type of modulation and the modulation upper frequency, defines the IF bandwidth necessary to handle the signal (remember to allow for drift). Therefore it is vital to have a knowledge of the more popular forms of signals in use today.

2.1 AMPLITUDE MODULATION

Amplitude modulation or AM, sometimes referred to as ancient modulation, has been the first practical form of voice transmission by radio waves and has been the workhorse of communications. AM is still heavily used today and should not be ignored, despite its inefficiency when compared to other more modern forms. AM is simply the amplitude variation of a carrier directly proportional to the magnitude of the intelligence to be sent. This modulation process is shown graphically in Fig. (2-1).

Mathematically, let the carrier be represented by

$$A \sin (\omega t + \phi) \qquad (2\text{-}1)$$

where

A is magnitude

ω is $2\pi F$, and F is the RF carrier frequency

ϕ is an arbitrary angle

The information to be transmitted is selected for this illustration to be a simple sinusoid represented by:

$$B \cos \mu t \qquad (2\text{-}2)$$

where

   B is magnitude

   μ is $2\pi f$, and f is the frequency of the sinusoid to be transmitted

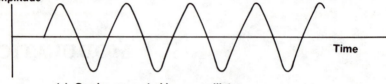

(a) Carrier generated by an oscillator.

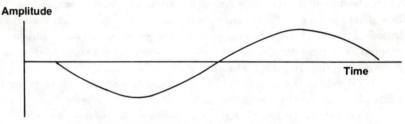

(b) Information to be transmitted.

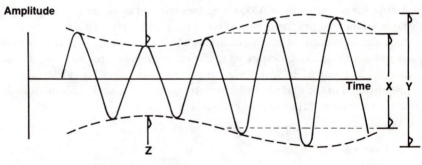

(c) Amplitude modulated carrier.

Fig. 2-1. Graphical representation of amplitude modulation.

Since *A* represents the magnitude of the carrier and this is to be varied or modulated in amplitude by $B \cos \mu t$ in a linear fashion, we add $B \cos \mu t$ to *A* and rewrite the carrier equation (2-1) to its amplitude modulated equivalent:

$$(A + B \cos \mu t) \sin (\omega t + \phi)$$

or

$$A (1 + B/A \cos \mu t) \sin (\omega t + \phi) \tag{2-3}$$

The ratio of B/A represents the magnitude of the modulation or modulation factor m. The value of m cannot ever exceed 1 and still retain a sinusoidal carrier form. Should $m > 1$, then the carrier would be interrupted during part of the modulation cycle and over modulation results, together with the generation of harmonics, which result in undesired spectral broadening known as splatter. This splatter causes interference, particularly to adjacent channel signals.

The percent of amplitude modulation is $(m \cdot 100)$, and is usually measured by the use of:

modulation meters
spectrum analyzers
oscilloscopes

The modulation meter is a calibrated receiver which uses a meter directly to indicate the percentage modulation.

Expanding (2-3) we have:

$$A (1 + m \cos \mu t) \sin (\omega t + \phi)$$

$$= A \sin (\omega t + \phi) + Am \cos \mu t \sin (\omega t + \phi)$$

$$= A \sin (2\pi F t + \phi) + \frac{Am}{2} \sin [2\pi (F + f) t + \phi]$$

$$+ \frac{Am}{2} \sin [2\pi (F - f) t + \phi] \qquad (2\text{-}4)$$

From these three terms the original carrier plus two additional frequencies symmetrically spaced about the carrier by f are found. These additional terms result from the modulation process and are referred to as sidebands. The maximum magnitude of these sidebands is found by letting $m = 1$ and is found to be $1/2$ that of the carrier.

The total width of the AM signal for this case is $2f$ and is seen to be independent of the modulation percentage of carrier power. Thus when a system is specified to contain frequencies to 10 kHz, the signals spectral width is 20 kHz.

Using a spectrum analyzer or selective voltmeter in the frequency domain, the y-axis represents magnitude and the x-axis frequency. The AM signal is found to be as shown in Fig. (2-2) and (2-3).

Most spectrum analyzers display the y-axis in power and the interpretation of the display must be altered accordingly.

In the time domain, the display of the voltage amplitude *versus* time of the AM wave results in the display of Fig. (2-1c) presented earlier. The value of m may be found from:

$$m\% = \frac{Y - X}{X} \; 100 \; (\textit{upward modulation}) \qquad (2\text{-}5)$$

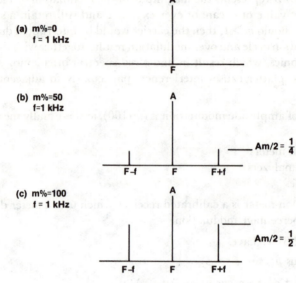

Fig. 2-2.
The AM signal displayed in the frequency domain where F is the carrier, A is magnitude, the modulating frequency is fixed, and $m\%$ is the variable.

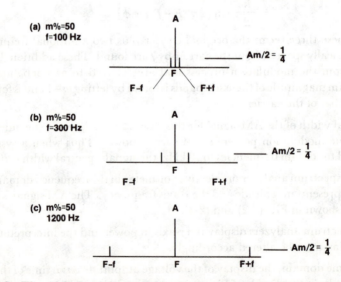

Fig. 2-3. The AM signal displayed in the frequency domain where F is the carrier, A is magnitude, m is constant, and the modulation frequency f is the variable.

An alternate form known as downward modulation is:

$$m\% = \frac{X-Z}{X}\ 100 \tag{2-6}$$

Complex modulating waveforms represented by $g(t)$ are treated in similar fashion as that of the sinusoidal case.

The AM wave may be described by:

$$A\,(1 + mgt)\,\sin\,(\omega t + \phi) \tag{2-7}$$

By representing $g(t)$ as a Fourier series and substituting this for $g(t)$, the resulting spectrum may be determined. This is shown graphically in Fig. (2-4). In this case the signal spectral width is $2f_3$. In practical complex cases, a filter is used to truncate the series, limiting the AM spectral width to a specified or practical value. The complex spectra is the summation of the AM spectra of each sinusoid contained in the series.

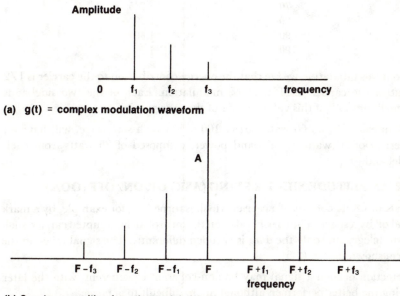

(a) g(t) = complex modulation waveform

(b) Spectrum resulting from the modulation of a carrier by complex wave g(t).

Fig. 2-4 (a) complex modulating wave spectrum, (b) the resulting AM spectrum.

The energy contained in the AM wave is the sum of that in the carrier plus in the side bands. From Eq. (2-4) it is seen that the energy in the carrier is unaffected by modulation. It is also seen that additional energy is added to the AM wave by the side bands. By squaring the magnitudes of Eq. (2-4), the energy in the wave may be computed for sinusoidal modulation as follows:

$$A^2 + 2 \left(\frac{Am}{2} \right)^2 = A^2 + \frac{2A^2 m^2}{4}$$

$$= A^2 \left(1 + \frac{m^2}{2} \right)$$

Table 2-1.
Relative AM Signal Energy *versus* %m

%m	Energy
0	1
10	1.005
20	1.02
30	1.045
40	1.08
50	1.125
60	1.18
70	1.245
80	1.32
90	1.405
100	1.5

From this tabulation we see that the energy contribution to the carrier is $1/2$ that of the carrier itself at 100% modulation. Each of the two sidebands contributes $1/2$ of this value or $1/4$ of the energy.

As an example, a 100 watt carrier 100% AM by a sine wave, will have an average of 50 watts of sideband power, composed of 25 watts from each sideband.

2.2 AMPLITUDE SHIFT KEYING (ASK) OR ON/OFF (OOK)

ASK or OOK consists of a carrier which is turned on, for example, by a mark and off by a space. The carrier takes on the form of an interrupted carrier, such as in telegraphy, only the data is encoded differently. The signal takes on the form shown in Fig. (2-5).

Detection of such a signal may be non-coherent or coherent, with the later being the better performer, although more difficult to achieve.

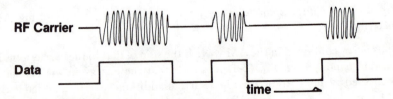

Fig. 2-5. The ASK signal in the time domain.

Non-Coherent Detection of ASK

Non-coherent detection in its simplest form consists of envelope detection followed by decision circuitry, as shown in Fig. (2-6). The decision threshold grossly affects the error probability (P_r) for mark and space independently, and they are therefore not equally probable. This results because the decision circuitry must distinguish between two signal states which are not equal in all respects. The mark or carrier on signal consists of carrier plus noise whereas the space or carrier off signal is noise alone. It has been shown (2-1) that P_r mark can be made equal to P_r *space for a given* $(C/N)_i$. A threshold of $\sim 50\%$ amplitude achieves this result [1].

Minimum probability of error results when a threshold of roughly

$$\left(\frac{1}{2} \; pulse \; amplitude \right) \cdot \left(1 + 2 \; \frac{e_b}{N_o} \right)^{\frac{1}{2}} \quad \text{is used.} \tag{2-8}$$

Where

e_b is the pulse energy
N_o is the noise density per reference bandwidth

In general, ASK is a poor performer although it is used in non-critical applications.

For $e_b/N_o \gg 1$ and a decision threshold of half the pulse amplitude, the probability of error for space is:

$$P_{e(space)} = e^{\left(-\frac{e_b}{2N_o} \right)} \tag{2-9}$$

and for mark

$$P_{e(mark)} = \frac{1}{(2\pi \, e_b/N_o)^{\frac{1}{2}}} \cdot e^{\left(-\frac{e_b}{2N_o} \right)} \tag{2-10}$$

From this, it is seen that the majority of errors are spaces converted to marks.

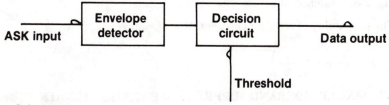

Fig. 2-6. Non-coherent detection of ASK.

Coherent Detection

Coherent detection requires a product detector with a reference signal which is phase coherent with the incoming signal carrier (see Fig. (2-7)).

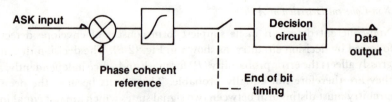

Fig. 2-7. Coherent detection of ASK using synchronous detection.

The product detector is followed by an integrator and a decision circuit timed to function at the end of 1 bit, or time τ.

An equivalent performer is the matched filter detector shown in Fig. (2-8). Here, the output of the matched filter is the convolution of the pulse and the impulse response of the matched filter. The resulting output is ideally diamond shaped and of duration $2\,\tau$, with a maximum signal energy at a time of $A^2/2$, where A is the signal amplitude and τ is the pulse duration. To complete the system, the decision circuitry is timed to function at time τ for optimum performance.

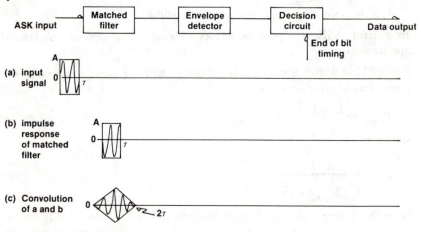

Fig. 2-8. Matched filter detection of ASK signals.

The probability of error for coherent ASK signaling is:

$$P_e = \frac{1}{2}\ \mathrm{erfc}\left(\frac{e_b}{2\mathcal{N}_o}\right)^{\frac{1}{2}} \qquad (2\text{-}11)$$

2.3   SINGLE SIDEBAND SUPPRESSED CARRIER SIGNALS (SSB SC)

An AM signal consists of a carrier and two sidebands and is described by:

$$A\sin\left(2\pi Ft + \phi\right) + \frac{Am}{2}\ \sin\ \left[2\pi\left(F + \mu\right)t + \phi\right]$$

$$+ \frac{Am}{2}\ \sin\ \left[2\pi\left(F - \mu\right)t + \phi\right] \qquad (2\text{-}12)$$

for a sinusoidal modulation signal of frequency μ and modulation factor m.

The carrier does not contain the modulation signal and its transmission serves only as a reference to the sidebands. The sideband components contain the modulating signal but are redundant. The single sideband principle utilizes these relationships, and removes the carrier and one of the sidebands from the AM spectrum, transmitting only the remaining sideband. (Upper or lower sideband as desired.) At the receiver, the carrier is reinserted and the modulation is recovered.

The principal problem with SSB SC is the accuracy required in the reinsertion of the carrier, which must be approximately 20 to 80 Hz or less from the missing carrier frequency, and on the proper side of the sideband. For upper sideband signals the carrier is reinserted on the low side. For lower sideband signals the carrier is reinserted on the high side. Failure to do this results in modulation inversion and unintelligibility. Because of the accuracy required for carrier reinsertion, SSB SC is seldom used in motional environments where doppler shifts are uncontrolled or variable.

The advantages of SSB SC are the narrower spectral occupancy of the transmitted signal, the reduced receiver IF bandwidth, and the lower transmitted power of SSB (compared to AM) for equivalence in $(S + N)/N$ at the receiver output. For an equivalent signal to noise ratio, the SSB signal requires a peak envelope power equal to $1/2$ that of the AM carrier with 100% modulation. Many comparisons may be made at this point. Comparing total powers radiated by both methods we have: modulation = 100% with sinusoidal modulating signal.

AM

carrier	1 unit
upper sideband	1/4 unit
lower sideband	1/4 unit
Total	1.5 units

SSB

0.5 units peak envelope power

Power ratio = 3.0 or 4.77 dB. Note that at AM and SSB equivalence, the AM total sideband power equals that of the SSB signal (PEP). When there is no modulation in AM the carrier is present, while with SSB there is no SSB signal. This extreme case shows a marked improvement in power utilized. Thus, where small, compact, efficient, low cost equipment is called for, SSB could be the answer. Since SSB SC signals are amplitude varying, they must be amplified linearly at the transmitter and receiver.

2.4 PULSE MODULATION

When a carrier is turned on and off, a pulse of RF energy is generated. If this is done repetitively at some rate such as in radar applications, digital data link, amplitude shift keying (ASK), or on/off keying (OOK), this energy takes on unique characteristics in the time and frequency domain. These characteristics are vital to the effective detection and processing of the signal and its contained data.

The detection of RF low level energy requires that noise be minimized while the signal is amplified. The noise component is kTB where B is bandwidth and is represented in terms of power by −144 dBm for a 1 kHz bandwidth. As B increases noise increases. Therefore, it becomes necessary to minimize the receiver bandwidth, to minimize the noise, and yet be of sufficient bandwidth to contain the majority of the signal energy.

The signal energy spectral occupancy and the receiver bandwidth can be computed by use of the Fourier transform on the time domain representation of the signal envelope.

For a pulse waveform this results in a spectral envelope described by sin x/x. More specifically, the spectrum is seen to consist of the repetition frequency and its harmonics; each with differing amplitude described by:

$$A_j = 2A \ \frac{T}{t_r} \left[\frac{\sin \pi j \ T/tr}{\pi j T/tr} \right] \hspace{3cm} (2\text{-}13)$$

where

$\quad$ A is amplitude

$\quad$ j is the harmonic of f_r

$\quad$ f_r is the repetition frequency

$\quad$ t_r is the repetition period $1/f_r$

$\quad$ T is the bit or pulsewidth

The generalized solution of A_j is shown in Fig. (2-9).

The energy of the signal is largely contained within the mainlobe of width $2/T$, which usually represents a sufficiently wide receiver RF and IF bandwidth, to efficiently process the pulse. Where pulse fidelity is important, a wider bandwidth may be necessary.

The detection bandwidth would be $1/2$ of the IF bandwidth with a spectral content on one side of that of the IF spectrum.

The spectrum, as seen on a spectrum analyzer, is unidirectional in the y or amplitude axis. All negative spectral lines are seen to be positive. When dealing with digital data, the bandwidth requirements should be based on a pulse that is one bit wide, for a worst case solution. The spectrum of ASK can be that of one bit, as a limit to all marks or 1's as the other; which is that of an unmodulated carrier. The bandwidth, of course, must accommodate the worst case.

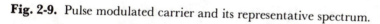

(a) Transmitted pulse envelope

(b) Envelope of $\frac{\sin x}{x}$; f_c is the carrier frequency.

(c) Spectral components of b

Fig. 2-9. Pulse modulated carrier and its representative spectrum.

2.4.1 Pulse Bandwidth Requirements

A pulse may be described by a series, consisting of harmonics of the repetition frequency. It follows that the pulse may be faithfully processed by a system with a bandwidth, which includes all of the terms of that series. In practice, system bandwidth is minimized, consistant with reasonable pulse fidelity. A pulse may be described by a $(\sin x)/x$ relationship. From this, the magnitude of the terms is seen to diminish in cyclic fashion, with increasing harmonic number. The inclusion of terms with minor magnitude, within a bandwidth window, can result in increased noise with little or no improvement in the signal. Therefore, it is necessary to restrict bandwidth to some practical value, determined by rise time or signal to noise ratio, or by a compromise of both.

Several relationships exist which help to define bandwidth requirements. If the pulses are treated as normalized Gaussian wave forms, the fractional basewidth of this wave form, within a specified time slot, may be related to bandwidth by the following relationships:

$$B = \left(\frac{2}{\pi\tau}\right) \left[2\ln\left(\frac{1}{k}\right)\right]^{\frac{1}{2}} \quad \text{(see [2])} \tag{2-14}$$

where

B is the system bandwidth

τ is the width of the time slot

k is the fractional base height of the normalized Gaussian pulse within the time slot

A further definition is shown in Fig. (2-10).

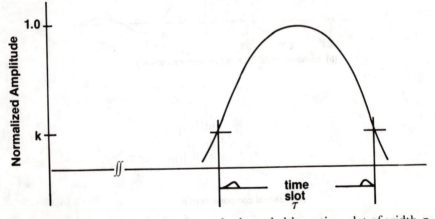

Fig. 2-10. A normalized Gaussian pulse bounded by a time slot of width τ where *k* is less than 1.

The bandwidth (B) computed contains 95.45% of the pulse energy. For pulses whose normalized magnitude is down to 0.1 within a 1 μsecond window, a bandwidth of 1.336 MHz is required as shown:

$$B = \left(\frac{2}{\pi\,10^{-6}}\right) \left[2\ln\left(\frac{1}{.1}\right)\right]^{\frac{1}{2}} = 1.336 \text{ MHz} \tag{2-15}$$

In digital systems the value of *k* determines the adjacent channel spillover and the lower the value of *k*, the better the fidelity of the system and the lower the spillover.

The rise time (t_r) of an ideal pulse applied to a band limited circuit may be approximated from:

$$t_r = \frac{0.35}{f_{max}} \tag{2-16}$$

where

t, is rise time

f_{max} is the upper −3 dB frequency response point of the video amplifier

For a f_{max} value of 1 MHz the pulse rise time is approximately 0.35 μsecond.

For the lower −3 dB response point of the video amplifier, the percentage tilt of a pulse train such as a 101010 may be approximated from:

$$Tilt\,(\%) = 100\,\pi\,\frac{f_{low}}{f} = 628.32\,f_{low}\,\tau \qquad (2\text{-}17)$$

where

f_{low} is the low frequency −3 dB point of the video amplifier

$f = 1/2\,\tau$ where τ is the bit width

Example: Let $\tau = 100$ μseconds

$f = 5$ kHz

For a tilt of 10%,

$$f_{low} = \frac{10f}{100\,\pi} = 0.3183f = 159 \text{ Hz}$$

Thus the frequency response of the video amplifier must extend to 159 Hz at the low end.

The reader is cautioned to realize that the receiver RF bandwidth must be twice the high frequency video bandwidth, for on/off keyed carrier signals.

2.5 FREQUENCY MODULATION

A carrier expressed by:

$A \sin (\omega t + \phi)$

or

$A \sin (2\pi Ft + \phi)$ \qquad (2-18)

can be described by letting the frequency of the carrier (F) vary linearly with the modulating frequency. The result of this is frequency modulation. Mathematically, this form is derived from a consideration of phase modulation.

Let $(2\pi Ft + \phi) = \delta$ \qquad (2-19)

and by definition the time rate of change of phase is frequency. Then:

$$\frac{d\delta}{dt} = \frac{d(2\pi Ft + \phi)}{dt} = \frac{d(2\pi Ft)}{dt} + \frac{d(\phi)}{dt} = 2\pi f$$

rewriting,

$$\frac{1}{2\pi}\,\frac{d\delta}{dt} = F \qquad (2\text{-}20)$$

Let F be varied or modulated by a sinusoid of the form:

$$\cos \mu t = \cos 2\pi ft \qquad (2\text{-}21)$$

Further, let us restrict the change in F by an amount ΔF.

We may write:

$$\frac{1}{2\pi} \frac{d\delta}{dt} = F + \Delta F \cos 2\pi ft \qquad (2\text{-}22)$$

Solving for δ by integration we have:

$$2\pi ft + \frac{\Delta F}{f} \sin 2\pi ft + \phi_{FM} \qquad (2\text{-}23)$$

Substituting into (2-18) we have the equation for FM.

$$A \sin \left(2\pi Ft + \frac{\Delta F}{f} \sin 2\pi ft + \phi_{FM} \right) \qquad (2\text{-}24)$$

Where $\Delta F \gg f$ then equation (2-24) reduces to the intuitive form:

$$A \sin (\omega t + \Delta F \sin \mu t + \phi_{FM}) \qquad (2\text{-}25)$$

The amount or degree of modulation is determined by the rate of the deviation ΔF to that specified for the equipment, or established by the Federal Communication Commission (FCC). The ratio of $\Delta F/f$ is termed the modulation index and is useful for the prediction of the spectral content of the FM signal. Conversely, the modulation index and deviation may be computed from the spectrum, for sinusoidal modulation. To establish the spectral prediction, equation (2-25) can be expanded as follows:

$$\frac{\Delta F}{f} = \beta \,, \ \omega = 2\pi F, \ and \ \mu = 2\pi f$$

$$A \sin (\omega t + \beta \sin \mu t + \phi_{FM}) \qquad (2\text{-}26)$$

let $\omega t + \phi_{FM} = \phi_1$

and $\beta \sin \mu t = \phi_2$

Since $\sin (\phi_1 \pm \phi_2) = \sin \phi_1 \cos \phi_2 \pm \cos \phi_1 \sin \phi_2$

Then we have:

$$A \left[\sin (\omega t + \phi_{FM}) \cos (\beta \sin \mu t) \right.$$
$$\left. + \cos (\omega t + \phi_{FM}) \sin (\beta \sin \mu t) \right] \qquad (2\text{-}27)$$

Using the relationships:

$$\cos (x \sin y) =$$

$$\mathcal{J}_o (x) + 2 \left[\mathcal{J}_2 (x) \cos 2y + \mathcal{J}_4 (x) \cos 4y + \mathcal{J}_6 (x) \cos 6y + \dots \right] \qquad (2\text{-}28)$$

and

$\sin (x \sin y) =$

$2 \left[\mathcal{J}_1 (x) \sin y + \mathcal{J}_3 (x) \sin 3y + \mathcal{J}_5 (x) \sin 5y + \mathcal{J}_7 (x) \sin 7y + \dots \right]$ (2-29)

where

$\mathcal{J}_n(x)$ are Bessel functions of the first kind and are identified by a capital $\mathcal{J}$

n is the order

x is the argument

Substituting and rewriting (2-27), we have:

$A \left[\sin (\omega t + \phi_{FM}) \right] \cdot \left[\mathcal{J}_o (\beta) + 2 \left[\mathcal{J}_2 (\beta) \cos 2 \mu t + \mathcal{J}_4 (\beta) \cos 4 \mu t + \dots \right] \right] +$

$A \left[\cos (\omega t + \phi_{FM}) \right] \cdot 2 \left[\mathcal{J}_1 (\beta) \sin \mu t + \mathcal{J}_3 (\beta) \sin 3 \mu t + \dots \right]$ (2-30)

Since

$$\sin p \cos q = \frac{1}{2} \sin (p + q) + \frac{1}{2} \sin (p - q) \qquad (2\text{-}31)$$

and

$$\cos p \sin q = \frac{1}{2} \sin (p + q) - \frac{1}{2} \sin (p - q) \qquad (2\text{-}32)$$

then,

$A \left[\mathcal{J}_o (\beta) \sin (\omega t + \phi_{FM}) + \right.$

$2 \left[\mathcal{J}_2 (\beta) \left(\frac{1}{2} \sin (\omega t + 2 \mu t) + \frac{1}{2} \sin (\omega t - 2 \mu t) \right) + \right.$

$\mathcal{J}_4 (\beta) \frac{1}{2} \sin (\omega t + 4 \mu t) + \frac{1}{2} \sin (\omega t - 4 \mu t) \quad + \dots$

$+ \mathcal{J}_1 (\beta) \left(\frac{1}{2} \sin (\omega t + \mu t) - \frac{1}{2} \sin (\omega t - \mu t) \right) +$

$\left. \left. \mathcal{J}_3 (\beta) \left(\frac{1}{2} \sin (\omega t + 3 \mu t) - \frac{1}{2} \sin (\omega t - 3 \mu t) \right) + \dots \right] \right]$ (2-33)

Rearranging in order:

$A \left\{ \mathcal{J}_o (\beta) \sin (\omega t + \phi_{FM}) \right.$

$+ \mathcal{J}_1 (\beta) \left[\sin (\omega t + \mu t) - \sin (\omega t - \mu t) \right]$

$+ \mathcal{J}_2 (\beta) \left[\sin (\omega t + 2 \mu t) + \sin (\omega t - 2 \mu t) \right]$

$+ \mathcal{J}_3 (\beta) \left[\sin (\omega t + 3 \mu t) - \sin (\omega t - 3 \mu t) \right]$

$+ \mathcal{J}_4 (\beta) \left[\sin (\omega t + 4 \mu t) - \sin (\omega t - 4 \mu t) \right]$

$\left. + \dots \dots \right\}$ (2-34)

In general form:

$$A \left\{ J_o\,(\beta) \sin\,(\omega t + \phi_{FM}) + \right.$$
$$\left. J_n\,(\beta)\,[\sin\,(\omega t + n\,\mu t) \pm \sin\,(\omega t - n\,\mu t)]\right\} \tag{2-35}$$

The sign in the J_n term is + for n even, and − for n odd.

For equation (2-34), it can be noted that there are sidebands for the fundamental modulation frequencies, as well as for the harmonics, spaced about the carrier. The magnitudes of these terms are easily determined from the Bessel charts or tables for small values of the modulation index.

Example 2-1: The deviation is given as 5 kHz at 1 kHz.

B = 5			
J	Magnitude (V)	V^2	dB
0	−0.177	.031	− 6.93
1	−0.327	.106	− 1.59
2	0.046	.002	−18.84
3	0.364	.132	− 0.641
4	0.391	.153	0
5	0.261	.068	− 3.52
6	0.131	.017	− 9.54
7	0.053	.0028	−17.37
8	0.018	.00032	−26.8
9	0.005	$25E^{-6}$	−37.87

2.5.1 Carson's IF Bandwidth Approximation Applied to FM Signal Spectral Requirements

Because the signal sidebands for FM follow the Bessel function curves of the first kind, it becomes requisite to determine the IF bandwidth necessary to include the significant spectral terms. A rule which works well for most cases was developed by J.R. Carson and is known appropriately as Carson's rule. This rule takes on the following representative forms:

$$B_{if} \approx 2\,(\Delta F + f_m),\ 1 \gg B \gg 1 \tag{2-36}$$

where

B_{if} is the IF bandwidth

ΔF is the peak deviation

f_m is the worst case (highest) modulating frequency

B is the modulation index

This equation is valid as shown for $1 \gg B \gg 1$. For those cases where this does not apply, a more suitable approximation is

$$B_{if} \approx 2 \, (\Delta F + 2f_m), \; 2 < B < 10 \tag{2-37}$$

2.5.2 Critical Determination of the IF Bandwidth Required for FM Signals by Power Summation

The FM process removes energy from a carrier and distributes it within the modulation sidebands. Modulation does not add energy to the signal as in AM. The distribution of this energy in the frequency domain is computable for simple modulation signals as shown in Section 2.5.

The individual sideband energy is proportional to the square of its Bessel coefficient and the sum of these energies for all sidebands, plus the carrier component is equal to that of the unmodulated carrier.

Then

$$\left[J_o \, (B) \right]^{2} + 2 \left[J_1 \, (B) \right]^{2} + 2 \left[J_2 \, (B) \right]^{2} + 2 \left[J_3 \, (B) \right]^{2} +$$

$$2 \left[J_n \, (B) \right]^{2} + \ldots \ldots = 1 \tag{2-38}$$

where

$$B = \frac{Peak \; Deviation}{Modulating \; Frequency}$$

Applying this equation to example (2-1), through the eight sidebands, it equals 99.324% of the total power.

Using this relationship it is possible to trade off IF bandwidth against signal to noise ratio, determining the penalties of signal energy loss. Complex modulations must be handled by computer because of the multiplicity of terms.

2.5.3 Foster Seeley FM Discriminator Detector (Fig. (2-11))

This popular FM discriminator detector consists of a driver-amplifier limiter, which feeds a constant amplitude FM signal to a specially-wound tuned IF transformer. This transformer consists of a tuned primary and secondary, with direct tap and mutual coupling. The result of this coupling is the generation of a secondary voltage, E_s, which is in quadrature to that of the primary voltage, E_P, at resonance. Two rectifier-type detectors are driven by the two outputs of the secondary. The outputs of the detectors are in opposition. At f_o the two vector inputs to the detectors are equal and opposite, resulting in no output, since $|E_a| - |E_b| = 0$. As the deviated carrier shifts frequency, the phase relationship is unbalanced and the detector drive vectors become unequal with the result $|E_a| - |E_b| \not\equiv 0$. The transfer characteristic of this circuit is the well-known S curve.

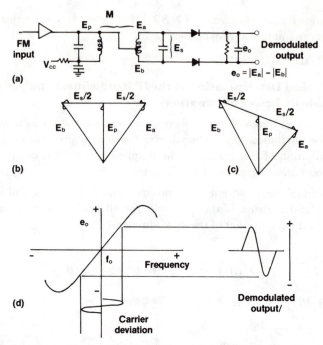

Fig. 2-11. (a) Foster Seeley discriminator; (b) vector relationships at center frequency (no deviation); (c) vector relationships with frequency offset F; (d) output transfer function.

2.5.4 Discriminator Detection of FM Signals Using Opposing AM Detectors

Since the carrier of an FM wave is frequency dependent upon the modulating wave and is of constant amplitude, it follows that detection can be achieved using two opposed AM demodulators tuned to different frequencies within the deviation of the carrier.

One implementation of this scheme is shown in Fig. (2-12).

The input FM signal is fed to two AM demodulators tuned to f_a and f_b where $f_b - f_a > 2\Delta F$ and ΔF is the peak deviation of the FM wave. The Q of the tuned circuits is such that an output is realized from each demodulator over $2\Delta F$. This results in each detector being a slope detector to the FM wave. Singly, the output of each detector has considerable distortion. By using two opposed frequency offset detectors, this distortion is largely removed. The resulting output is an S curve typical of discriminators.

2.5.5 Discriminator Output Filtering

The output signal to noise ratio of a limiter discriminator is described by [1]:

$$SNR_o = 3 \left(\frac{\Delta F}{f_m} \right)^2 \left(\frac{C}{2 N_o f_m} \right)$$

where

ΔF is the peak deviation (Hz)

f_m is the upper cutoff of the output lowpass filter

C is signal power

N_o is the one-sided noise power density in watts /Hz

The above equation is valid for IF signal to noise ratios > 10 dB where full FM improvement is realized.

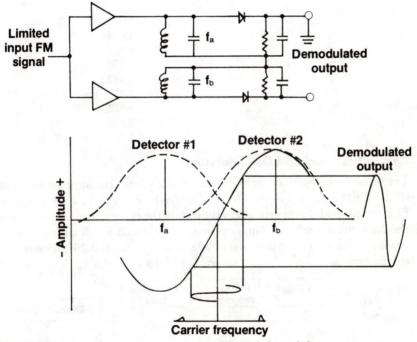

Fig. 2-12. FM detection using two opposed AM demodulators.

McKay has shown that by using a bandpass filter at the discriminator output, an improvement factor I results and is described by [3]:

$$1 = \frac{1}{1 - P^3}$$

where

$$P = \frac{f_{min}}{f_{max}}$$

represents the ratio of upper to lower cutoff frequency ratio of the ideal bandpass filter.

This improvement factor increases as the cube of P and is shown in Table 2-2. It is therefore desirable to limit the low frequency of the discriminator where such information is not present or useful.

Improvement Factor Resulting from Bandpass Filtering
the Output of a Discriminator Detector

Ratio of f_{min}/f_{max}	Improvement Factor
0	1
.1	1.001
.2	1.008
.3	1.027
.4	1.068
.5	1.142
.6	1.275
.7	1.522
.8	2.049
.9	3.69

2.5.6   Phase Lock FM Demodulation

The phase lock demodulator of Fig. (2-13) is a conventional phase lock loop (PLL) which is locked to the FM carrier. As the FM carrier is deviated, the PLL error signal is proportional to the shift of the carrier, and may be used as the demodulated output. The loop bandwidth must include all modulation terms of interest. This type of demodulator is readily available in integrated circuit form, and is readily adaptable for FM demodulator applications.

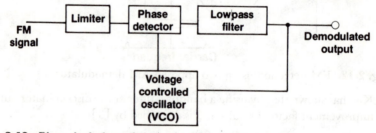

Fig. 2-13. Phase-lock demodulator for FM signals.

2.5.7   Counter FM Demodulator

The counter FM demodulator of Fig. (2-14) is noted for its wide bandwidth and excellent linearity. Unfortunately it is useful only for large deviation applications because of its low sensitivity. For those applications it is excellent.

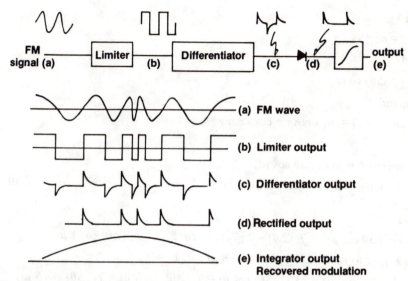

Fig. 2-14. Counter-FM demodulator.

Through limiting, the input FM signal results in a flat-topped waveform, resembling a square wave. This wave form is differentiated usually by an RC network, and rectified to preserve only one polarity of the differentiated waveform. Integration of this signal provides the demodulated output.

2.6 PHASE MODULATION

In phase modulation, the phase of a carrier is varied with a modulation signal.

A carrier of the form:

$$A \sin (\omega t + \phi)$$

can be phase modulated by adding a phase variable controlled by a modulating signal. This phase variable must be limited to a maximum value $\Delta\theta$. For a simple sinusoidal modulating signal this phase variable becomes:

$$\Delta\theta \cos \mu t$$

where

$$\mu = 2\pi f$$

f is the modulating frequency

$\Delta\theta$ is in radians

Including this term, we have the phase modulated wave described by:

$$A \sin (\omega t + \Delta\theta \cos \mu t + \phi_o) \tag{2-39}$$

Let

$$x = \omega t + \phi_o$$

and

$$y = \Delta\theta \cos \mu t$$

Expand

$$\sin (x + y) = \sin x \cos y + \cos x \sin y$$

or

$$\sin (\omega t + \phi_o) \cos (\Delta\theta \cos \mu t)$$
$$+ \cos (\omega t + \phi_o) \sin (\Delta\theta \cos \mu t) \qquad (2\text{-}40)$$

and

$$\cos (\Delta\theta \cos \mu t) = J_o (\Delta\theta) - 2 [J_2 (\Delta\theta) \cos 2 \mu t - F_4 (\Delta\theta) \cos 4 \mu t$$
$$+ J_6 (\Delta\theta) \cos 6 \mu t - J_8 (\Delta\theta) \cos 8 \mu t \dots]$$
$$\sin (\Delta\theta \cos \mu t) = 2 [J_1 (\Delta\theta) \cos \mu t - J_3 (\Delta\theta) \cos 3 \mu t + J_5 (\Delta\theta) \cos 5 \mu t$$
$$- J_7 (\Delta\theta) \cos 7 \mu t + \dots] \qquad (2\text{-}41)$$

Then

$$A [\sin (\omega t + \phi_o) [J_o (\Delta\theta) - 2 [J_2 (\Delta\theta) \cos 2 \mu t - J_4 (\Delta\theta) \cos 4 \mu t + \dots]]$$
$$+ \cos (\omega t + \phi_o) 2 [J_1 (\Delta\theta) \cos \mu t - J_3 (\Delta\theta) \cos 3 \mu t + \dots]] \qquad (2\text{-}42)$$

Where $J_n (\Delta\theta)$ are Bessel functions of the first kind and $(\Delta\theta)$ is the argument given in radians.

Rearranging

$$A [J_o (\Delta\theta) \sin (\omega t + \phi_o)$$
$$+ J_1 (\Delta\theta) \cos (\omega t + \mu t + \phi_o) + J_1 (\Delta\theta) \cos (\omega t - \mu t + \phi_o)$$
$$- J_2 (\Delta\theta) \sin (\omega t + 2 \mu t + \phi_o) - J_2 (\Delta\theta) \sin (\omega t - 2 \mu t + \phi_o)$$
$$- J_3 (\Delta\theta) \cos (\omega t + 3 \mu t + \phi_o) - J_3 (\Delta\theta) \cos (\omega t - 3 \mu t + \phi_o)$$
$$+ J_4 (\Delta\theta) \sin (\omega t + 4 \mu t + \phi_o) + J_4 (\Delta\theta) \sin (\omega t - 4 \mu t + \phi_o)$$
$$+ J_5 (\Delta\theta) \dots] \qquad (2\text{-}43)$$

The phase modulation process does not add power to the signal but redistributes the carrier energy in the form of sidebands. In other words, the sum of the powers in the spectrum is equal to that of the unmodulated carrier.

For a fixed $\Delta\theta$, the respective sidebands are fixed in magnitude. As the modulating frequency is allowed to approach 0, the width of the spectrum collapses to 0. Thus, for phase modulation the spectral width is directly proportional to the modulating frequency.

2.6.1 Carson's IF Bandwidth Approximation Applied to PM Signal Spectral Requirements

Carson's rule for FM may be applied to PM signals by slightly modifing equations (2-36) and (2-37). This is done by multiplying by f_m/f_m and substituting M (the modulation index for PM) for $\Delta F/f_m$ (the modulation index for FM). Thus, the IF bandwidth required to accomodate the significant spectral sidebands for PM is:

$$B_{if} = 2\,(M+1)\,f_m,\ 1 \gg M \gg 1 \tag{2-44}$$

and

$$B_{if} = 2\,(M+2)\,f_m,\ 2 < M < 10 \tag{2-45}$$

where

M is the modulation index for PM

f_m is the worst case highest modulating frequency

2.6.2 PM Signal IF Bandwidth Requirements Using Power Summation of the Side Bands

This determination is identical to that of FM and the reader is referred to section 2.5.2. The only difference is the argument of the Bessel function, where the peak phase deviation in radians is substituted for the FM index B.

2.7 PHASE SHIFT KEYED SIGNALS (PSK)

As implied by the title, the digital modulation is impressed upon the carrier in such a manner that the phase of the carrier is shifted for a mark or space by a different fixed amount each time either occurs.

In the simplest form, a mark may cause a shift of π radians, while a space may not shift the carrier at all. Thus, the digital modulation is transmitted as a series of 0 or π shifts in carrier phase. Such waveforms are easily generated using a common doubly balanced mixer, by applying the digital data to the IF port and the carrier to the LO port. The PSK signal then appears at the RF port. This signal has the spectral form of $(\sin x)/x$, or more specifically

$$F_{(\omega)} = A\tau\ \frac{\sin\left[(\omega_l - \omega_o)\dfrac{\tau}{2} + \phi\right]}{\left[(\omega_l - \omega_o)\dfrac{\tau}{2} + \theta\right]} \tag{2-46}$$

where

A is the amplitude of the signal

τ is the bit width

ω is the carrier radian frequency

θ is an arbitrary phase

The resulting spectrum and receiver IF signal has the spectral form shown in Fig. (2-15).

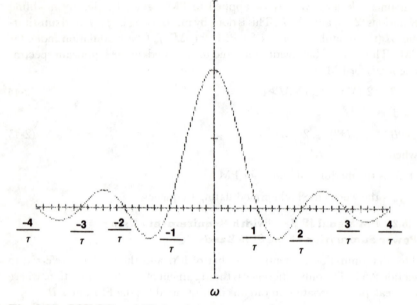

Fig. 2-15. PSK RF and IF signal spectrum envelope.

The majority of the signal is contained in the mainlobe and because of the masking of the minor lobes by noise near the detection threshold, the IF bandwidth of the receiver seldom exceeds $(2 \text{ or } 3)/\tau$. And the post-detection bandwidth is $1/2$ of the IF bandwidth.

There is one important spectral consideration and that is for a symmetrical square wave modulation. For example in a 10101010.... data pattern, the carrier is suppressed. In fact for random patterns, this suppression will vary and as a limit will approach the square wave case.

The phase states in PSK, need not be limited to 2, in fact it is feasible to utilize 2^n discrete states for n commonly up to 4 (16 phase). Where $n = 1$, the signal is referred to as biphase (BPSK) and quadriphase where $n = 2$ (QPSK).

As n increases, the channel can handle more information, but at a sacrifice in noise immunity. The number of separate data channels a system can handle is $2^n/2$. Therefore, BPSK can handle one data stream, QPSK can handle two data streams and eight phase can handle four data streams *et cetera*.

The phase states are usually separated equally as follows, to reduce noise problems:

n	Phase	Separation (degrees)
1	2	180
2	4	90
3	8	45
4	16	22.5

The phase relationships for a QPSK system are shown in Fig. (2-16).

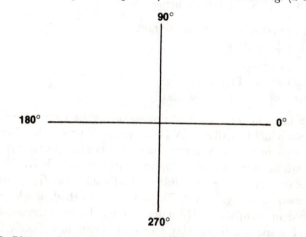

Fig. 2-16. Phase relationships for a QPSK system, showing a total of four phase states for a two-signal data capability.

Other angular relationships are usable and often preferred because diametrically opposed phase data results in carrier nulls, making carrier recovery more difficult. The demodulator process requires a coherent phase reference because the data is represented as changes in carrier phase. This necessitates the use of coherent or product detection techniques. A product detector is a three terminal device. It has a signal port, a signal carrier reference port, and an output port. Any form of mixer, phase detector, multiplier *et cetera* qualifies as the detector. The detector is shown in block form in Fig. (2-17).

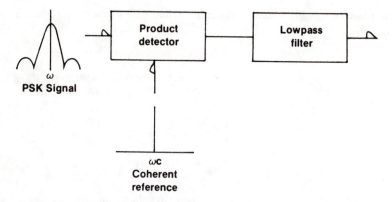

Fig. 2-17. Basic PSK detection scheme.

There are three popular detection methods:

Coherent (Fig. (2-17))
Differentially coherent
Delay line method (Fig. (2-18))
Multiplier loop reference (Fig. (2-19))

Of these, the later two methods are the most popular. Coherent techniques offer an improvement of 1 to 3 dB in e_b/N_o for a given error probability, but are more difficult to implement, since a carrier phase reference is required at the receiver. The differentially coherent and multiplier loop techniques derive the reference carrier from the signal. The delay line method is shown in Fig. (2-18) and the multiplier method in Fig. (2-19). The delay line method, also known as the Kineplex system, compares the 1 bit delayed signal against the signal without any delay. The use of a fixed delay limits the system to a specific bit width, restricting its applications. It also requires differential encoding at the transmitter.

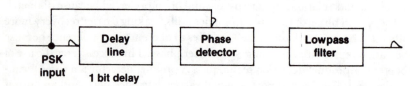

Fig. 2-18. Differentially coherent detection of PSK signals using a delay of 1 bit.

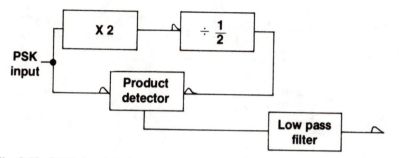

Fig. 2-19. PSK signal detection using a multiplier loop to recover the carrier for biphase modulation systems.

The multiplier technique generates the carrier by multiplying the BPSK carrier shift of zero and π by two, resulting in shifts of zero and $2\pi = 0$. This strips all modulation and leaves a carrier at twice the frequency. The on frequency carrier is recovered by dividing by two. This technique does not have the limitations of the delay line method and may be used with any data rate within its system bandwidth.

The delay line method cannot process data without differential encoding at the transmitter. The encoding system is shown in Table 2-3 and Fig. (2-20).

Table 2-3.
Differential Encoding DPSK

Transmitter

Message		0	1	1	1	0	1	0	0	0	1	1	
Encoding	1*	0	0	0	0	1	1	0	1	0	0	0	
Transmitter phase	π	0	0	0	0	π	π	0	π	0	0	0	

Receiver

Received phase	π	0	0	0	0	π	π	0	π	0	0	0	
1 bit delay	–	π	0	0	0	0	π	π	0	π	0	0	0
Phase detector	–	0	1	1	1	0	1	0	0	0	1	1	–

*Arbitrary start bit

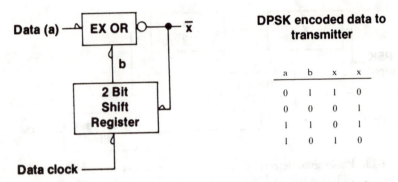

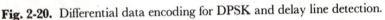

Fig. 2-20. Differential data encoding for DPSK and delay line detection.

The data is fed into an exclusive NOR circuit, which has as the second input its own output, shifted by 1 bit. The truth table shows that when inputs *a* and *b* are alike one output results, and conversely, when they differ, a zero becomes the output.

Signal derived references are corrupted by noise at low signal levels, resulting in performance degradation. Of the three systems, the performance rating, based on error probability for a given e_b/N_o, in decending order, are: coherent, multiplier, and delay line. However, the performance difference of all three systems is within a 1 to 3 dB window.

The probability of error for the three systems has been defined as follows:

Coherent PSK

$$P_e = \frac{1}{2} \ \mathrm{erfc} \ \sqrt{\frac{e_b}{N_o}} \qquad\qquad (2\text{-}47)$$

Signal Derived Reference PSK

$$P_e = \frac{1}{2} \ e^{-e_b/N_o} \qquad\qquad (2\text{-}48)$$

At large e_b/N_o the performance is less than 1 dB worse than coherent PSK. For $P_e = 10^{-1}$ the curves are 3 dB apart, in favor of coherent detection.

2.8 FREQUENCY SHIFT KEYING (FSK)

In FSK signaling, binary data causes the transmission of one frequency for mark and a different frequency for space. These frequencies can originate from two discrete oscillators or from a single oscillator, which is frequency modulated by the two-state digital modulation. These two methods are shown in Fig. (2-21).

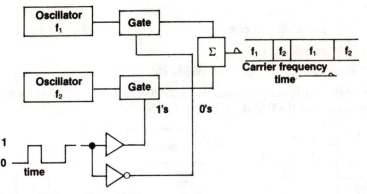

Fig. 2-21a. Two-oscillator FSK transmission.

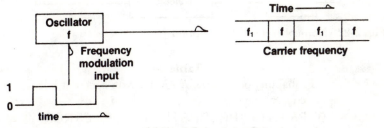

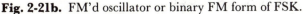

Fig. 2-21b. FM'd oscillator or binary FM form of FSK.

Reception of FSK has two possible solutions which are:

    non-coherent detection

    coherent detection

It will be shown that coherent detection is superior (although seldom used) because it requires a coherent reference signal in the detection process.

Non-coherent detection systems are easily implemented and consist of two detection channels. Each channel utilizes a bandpass filter and an ordinary AM envelope detector. One filter is tuned to the mark frequency and the other to the space frequency. The detector outputs differ in polarity and are summed to produce the output. This configuration is shown in Fig. (2-22). Each channel is in reality a single on/off carrier (OOK) system, both of which take turns as dicatated by the modulation.

The performance of the noncoherent system of FSK reception is given by:

$$P_e = \frac{1}{2} \exp\left[-\frac{1}{2}\left(\frac{e_b}{\mathcal{N}_o}\right)\right] \qquad (2\text{-}49)$$

where

P_e is the probability of error

e_b is the energy per bit

N_o is the noise density per unit bandwidth

The solution of P_e may be found using (2-49) and a series of solutions may be secured through the BASIC program of Table 2-4.

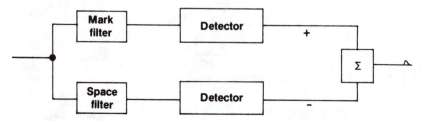

Fig. 2-22. Non-coherent FSK reception.

Table 2-4.

Probability of Error *vs.* E_b/N_o for Non-coherent FSK

```
10 SHORT P1
20 PRINT "PROBABILITY OF ERROR
      VS EB/NO FOR FSK"
30 PRINT
40 PRINT "*************************
      ************"
50 PRINT
60 PRINT "EB/NO";TAB(18);"PE"
70 DISP "ENTER EB/NO,(DB), MIN,
      MAX,STEP"
80 INPUT A1,B1,S1
90 FOR K=A1 TO B1 STEP S1
100 K1=10^(K/10)
110 P1=.5*EXP(-(.5*K1))
120 PRINT K;TAB(18);P1
130 NEXT K
140 END
```

There are several variations of the detection scheme which offer improvement. These include weighing the two detector outputs and making a decision based upon which detector has the largest output. For this variation, both detectors have like polarity outputs and are fed into a differential comparator. Another variation involves the use of discriminator detection of the FSK signal.

2.8.1   FSK Signal Spectrum and Bandwidth Considerations

FSK consists of pulses of RF energy alternating between two frequencies such that RF output is present at one frequency or the other. The duration of one such pulse or bit possesses the characteristics typical of any, and also represents the widest spectral occupancy.

For an example let a mark bit be represented by:

$f(t) = A \sin(\omega_l t + \theta)$

where

$$-\tau/2 < t < \tau/2 \qquad\qquad (2\text{-}50)$$

and

  A is magnitude

  ω_1 is $2\pi \cdot$ (mark frequency)

  θ is an arbitrary phase angle

  τ is the bit width

In the frequency domain this becomes:

$$F(\omega) - A\,\tau/2 \left[e^{j(\theta - \pi/2)} \cdot \frac{\sin(\omega - \omega_l)\,\tau/2}{(\omega - \omega_l)\,\tau/2} \right.$$

$$\left. + e^{-j(\theta - \pi/2)} \cdot \frac{\sin(\omega + \omega_l)\,\tau/2}{(\omega + \omega_l)\,\tau/2} \right] \qquad (2\text{-}51)$$

Thus the spectrum is seen to have a $(\sin x)/x$ response as depicted below:

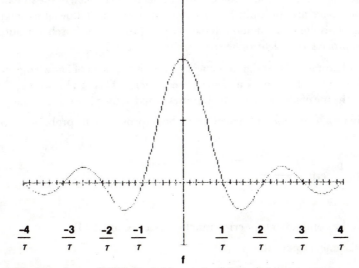

Fig. 2-23. Envelope of FSK for 1 bit mark of duration τ.

There exists a second such spectrum at the space frequency. The total spectral occupancy of the signal is the sum of the two. In many cases the mark and space spectrums partially overlap reducing the total bandwidth.

It can be noted from Fig. (2-23) that the majority of the signal energy is contained in a frequency bandwidth of $2/\tau$ MHz, (where frequency is the reciprocal of time). The total spectrum of the FSK signal includes two spectra separated by a guard frequency. In the interest of spectral conservation and receiver carrier to noise ratio, the guard frequency is made necessarily small, resulting in spillover of mark and space energy into each others filter bandwidths. As a rule, the guard band is made equal to $2/\tau$ MHz, using mark and space bits of τ seconds.

It has been shown that the mark and space filter bandwidths of $1.5/\tau$ are good choice from a noise and intersymbol interference standpoint [4].

The RF and IF bandwidths are approximately:

$$B_{if} \approx 2\,(D + f_m) \tag{2-52}$$

where

D is the shift of the carrier from its mean value, and

f_m is $1/\tau$

2.8.2   Coherent FSK

The optimum demodulator utilizes matched filtering. This can be realized by utilizing product detection (coherent multipliers) where each of the signals for mark and space are multiplied by a coherent reference, followed by lowpass filtering. This coherent demodulated output produces a baseband output, which is processed to determine the dominant output.

The disadvantage of this signal recovery mean is the need of knowledge of the exact frequency of the mark and space frequencies. This can be accomplished through the use of phase-locked techniques tuned to the respective frequencies.

The advantage of coherent detection is the improved error probability given by:

$$P_e = \frac{1}{2}\ \mathrm{erfc}\left(\frac{e_b}{2\,N_o}\right)^{\frac{1}{2}}$$

where

$erfc$ is the complimentary error function (see section 4.14)

e_b is the energy per bit

N_o is the noise density

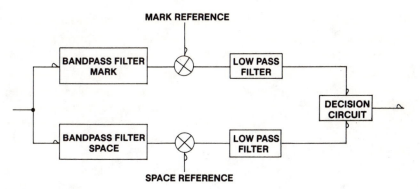

Fig. 2-24. Coherent FSK demodulator block diagram.

This form of demodulator is shown in block form in Fig. (2-24). Because of the additional complexity, this form of signal recovery is seldom utilized except in critical cases.

REFERENCES

[1] Panter, P.F., *Modulation and Noise and Spectral Analysis.* New York: McGraw-Hill, 1965.

[2] Noel, F. and Kolodzey, J., "Nomograph Shows Bandwidth for Specified Pulse Shape," *Electronics*, April 1, 1976, p. 102.

[3] Kay, G.A., "Signal to Noise Ratio in the Band Pass Output of a Discriminator," *IEEE Transactions on Aerospace and Electronic Systems,* vol. AES-6, no. 3, p. 340.

[4] Farone, J.N., and J.R. Feldman, *An Analysis of Frequency Shift Keying Systems.* Armour Research Foundation Technology Center, Chicago, Illinois.

3

NOISE

3.1 EXTERNAL NOISE

A communications link is affected by external noise which degenerates the performance of that link. It is not fruitful to have a low noise receiver capable of signal reception of −115 dBm signals in the screen room when it is later placed in service where the external noise level is −90 dBm. In such a situation a less costly receiver design could suffice, but the link specifications would not be met except in the screen room. In such a situation either more transmitting power, higher antenna gain, or both are the only answer, short of relocation.

There are several sources of noise, including:

Galactic origination outside the earth's atmosphere
Atmospheric disturbance and storms
Manmade

The characteristics of these noise sources are shown graphically as a function of frequency in Fig. 3-1. The average noise intensity is shown on a relative basis referenced to kTB. Considerable global and seasonal variance exists for these characteristics, and serious concern for these noise sources should involve a specific site investigation. Generally, it can be concluded that noise diminishes with increasing frequency and distance from urban environments.

External noise may be referred to the antenna effective temperature by:

$$T_a = T_o f_a \qquad (3\text{-}1)$$

where

$T_o = 290°K$
f_a is the effective antenna noise factor

and

$$f_a = \frac{P_n}{k T_o B}$$ (3-2)

where

P_n is the noise power at the antenna, loss-free
$k = 1.38 \cdot 10^{-23}$ joules/°Kelvin
$T_o = 290°$ Kelvin
B is the bandwidth in Hz

See reference [1] for more information.

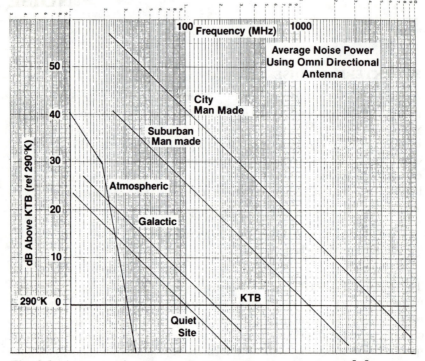

Fig. 3-1. Average noise power using omni-directional antenna [1].

3.2 EFFECTIVE NOISE BANDWIDTH

A filter response is often difficult to describe mathematically because of its non-ideal characteristic. A more useful convention results when the filter is idealized by an equivalent rectangular shape of equal area. This is done by integrating the power response of the filter or circuit. Do not confuse this with dB or voltage responses, or erroneous results will be obtained. Next, compute an equivalent area rectangle whose magnitude is equal to the original response. These two areas, being equal, result in an ideal equivalent bandpass. The width of this rectangle is the equivalent noise bandwidth. This is also sometimes referred to as ENB, or simply, noise or ideal bandwidth. As an illustration, a much used application is the determination of noise power after a filter. Here, the noise is usually white with a given spectral density (D) in convenient units such as power in milliwatts or dBm per unit bandwidth. Multiplying the noise spectral density by the equivalent noise bandwidth gives the noise power.

The response of Fig. (3-2a) is readily measured and integrated by using: the polar planimeter, the rectangular, the trapazoidal, or Simpson's rule. Converting to an equivalent height rectangle gives (3-2b), which is the idealized equivalent noise bandwidth $(f_b - f_a)$ where the heights and areas of both are equal.

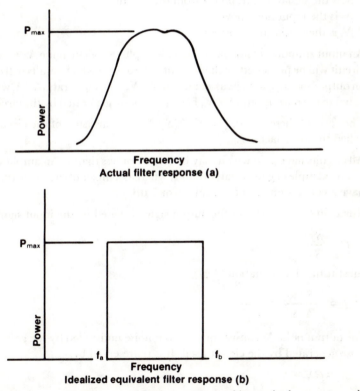

Fig. 3-2. An example of an actual filter response and its equivalent rectangular bandwidth or effective noise bandwidth. Both responses are of equal height and area.

3.3 NOISE FIGURE AND NOISE FACTOR

By definition noise figure is noise factor expressed in dB.

$$NF = 10 \log_{10} F \tag{3-3}$$

where

NF is noise figure
F is noise factor

Furthermore,

$$F = \frac{S_i / N_i}{S_o / N_o} \tag{3-4}$$

where

S_i is signal input power applied to the circuit
S_o is the signal output power from the circuit
N_i is the input noise power
N_o is the output noise power

An input stimulus, consisting of a signal S_i plus associate noise N_i, applied to a circuit will be processed by that circuit. The output which results will consist of an output signal S_o with its associated noise N_o. The input ratio S_i / N_i will not be equal to the output ratio S_o / N_o, because of noise generated by the circuit itself.

The ratio of these ratios, $(S_i / N_i) / (S_o / N_o)$, is a measure of the circuit and is called the noise factor F.

All circuits have gain which may be greater or less than 1. An amplifier would be an example of gain greater than 1, whereas a mixer of the diode type would have a conversion loss of typically 6 or 7 dB.

The gain G is the ratio of the output signal divided by the input signal. Or:

$$G = \frac{S_o}{S_i} \tag{3-5}$$

Substituting into equation (3-4)

$$F = \frac{S_i N_o}{N_i S_o} = \frac{N_o}{G N_i} \tag{3-6}$$

The output noise N_o consists of the input noise multiplied by the gain plus the noise generated by the circuit itself. It is represented by:

$$N_o = G N_i + N_x \tag{3-7}$$

where

N_x is the noise generated internally in the circuit appearing at its output terminals

Then

$$F = \frac{GN_i + N_x}{GN_i} = 1 + \frac{N_x}{GN_i} \qquad (3\text{-}8)$$

and

$$N_i = kTB \qquad (3\text{-}9)$$

where

k is Boltzmann's constant and is equal to $(1.38044 \pm 0.00007) \cdot 10^{-23}$ joules /° Kelvin

T is the source temperature, usually taken as 290° Kelvin

B is the effective noise bandwidth (Hz)

Equation (3-8) becomes:

$$F = 1 + \frac{N_x}{GkTB} \qquad (3\text{-}10)$$

Solving for the circuit noise N_x we have:

$$N_x = (F - 1) GkTB \qquad (3\text{-}11)$$

The value of B is the equivalent rectangular bandwidth of area equal to that of the device.

From equation (3-6)

$$F = \frac{N_o}{GN_i} \qquad (3\text{-}12)$$

then

$$N_i F = \frac{N_o}{G} \qquad (3\text{-}13)$$

Since the output noise of a circuit divided by gain must be the equivalent noise input to the circuit, then $N_i F$ is the equivalent circuit, input noise power. From this we have the important relationship:

$$N_{i\,equiv.} = N_i F = kTBF \qquad (3\text{-}14)$$

In receiving applications, kTBF represents the noise power at the input of the receiver, which the signal level must overcome. If the input signal power is equal to kTBF then $S/N = 1$. For computation ease, the log form of kTBF is suggested as follows:

$$10 \log kT = -204 \text{ dBw} = -174 \text{ dBm} \qquad (3\text{-}15)$$

where

dBw refers to dB relative to 1 watt

and

dBm to dB relative to 1 milliwatt

Then

$$10 \log kTB = -174 + 10 \log B, \text{dBm}$$
$$= -174 \text{ dBm/Hz}$$
$$= -144 \text{ dBm/kHz}$$
$$= -114 \text{ dBm/MHz} \qquad (3\text{-}16)$$

This readily shows the importance of minimizing bandwidth to reduce noise. Fig. (3-3) relates kTB in dBm to bandwidth B for quick reference and is of sufficient accuracy for most computations.

Example: It is specified that a receiver must process a −97 dBm unmodulated carrier with a S/N of 10 dB. The receiver bandwidth is given as 50 kHz. What is the maximum noise figure the receiver can have?

From Fig. (3-3) kTB = −127 dBm
Then $NF = 127 - 97 - 10 = 20$ dB
A noise figure of 20 dB is the maximum allowable value.

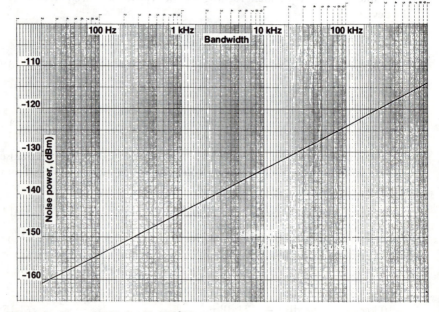

Fig. 3-3. Plot of kTB for T=290°K.

3.4 TEMPERATURE

Noise temperature is related to noise factor (F). This relationship may be derived by modeling an actual amplifier by using an ideal (noiseless) amplifier preceded by a summing junction with two inputs. The two inputs are the input noise and the noise generated within the real amplifier. This is shown below:

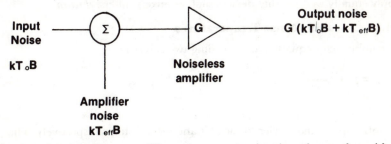

Fig. 3-4. Model of an amplifier using a summing junction and an ideal amplifier plus two noise sources.

$$input\ noise\ =\ kTB:\ amplifier\ noise\ =\ kT_{eff}B \qquad (3\text{-}17)$$

where

k is Boltzmann's constant $(1.38044 \pm 0.00007) \cdot 10^{-27}$ 7 watts/° Kelvin
T_o is 290° Kelvin (unless specified otherwise)
B is bandwidth in Hz
T_{eff} is the effective noise temperature of the amplifier

The amplifier has a gain G, therefore the output noise is:

$$\mathcal{N}_o = G\,(kT_oB + kT_{eff}B) \qquad (3\text{-}18)$$

Since noise factor F is the ratio of signal to noise in, to signal to noise out, or

$$F = \frac{\dfrac{S_i}{\mathcal{N}_i}}{\dfrac{S_o}{\mathcal{N}_o}} = \frac{S_i\mathcal{N}_o}{S_o\mathcal{N}_i} = \frac{1}{G}\,\frac{\mathcal{N}_o}{\mathcal{N}_i} \qquad (3\text{-}19)$$

$$F = \frac{1}{G}\,\frac{(kT_oB + kT_{eff}B)\,G}{kT_oB} \qquad (3\text{-}20)$$

$$F = 1 + \frac{T_{eff}}{T_o} \qquad (3\text{-}21)$$

Knowing the noise factor, the effective noise temperature may be found from:

$$T_o (F-1) = T_{eff} \tag{3-22}$$

Note that these relationships are gain independent.

Although an amplifier was used in these derivations, the resulting equations apply equally well to other devices such as mixers, filters, *et cetera*.

3.4.1 Cascading Noise Temperatures

The noise factor equation for cascading two stages is:

$$F_t = F_1 + \frac{F_2 - 1}{G_1} \tag{3-23}$$

where

Subscripts 1 and 2 refer to the first and second stages respectively. The first stage is the input stage.

Substituting equation (3-21) into the above

$$F_t = 1 + \frac{T_{eff_t}}{T_o} = 1 + \frac{T_{eff1}}{T_o} + 1 + \frac{\dfrac{T_{eff2}}{T_o} - 1}{G_1}$$

or,

$$T_{eff_t} = T_{eff1} + \frac{T_{eff2}}{G_1} \tag{3-24}$$

Example: A filter with a 3 dB loss (and a 3 dB noise figure) is placed ahead of an amplifier with an effective noise temperature of 864°K. What is the new overall noise temperature?

The 3 dB noise figure is converted to effective noise temperature.

$$290(1.995 - 1) = 288.55°K$$

then

$$T_{eff_t} = 288.55 + (864/.501) = 2013°K$$

This is a noise factor of (using equation (3-21))

$$F = 1 + (2013/290) = 7.94$$

The noise figure is:

$$NF = 10 \log_{10} 7.94 = 8.998 \text{ dB}$$

Where more than two stages are involved the stages are taken two at a time and the process is repeated as often as necessary. This piecewise solution avoids lengthy equations and allows stage by stage examination.

3.5 CONVERSION NOISE

3.5.1 Ideal Noise Free Signal and Local Oscillator Signals

The ouput noise from a mixer is found from the definition of noise factor.
From

$$F = \frac{S_i N_o}{S_o N_i} = \frac{N_o}{GN_i} = \frac{N_o}{kTBG} \tag{3-25}$$

where

F is noise factor
S_i is input signal
S_o is output signal
kTB is the noise in a bandwidth B
G is gain

The signal output of the mixer is:

$$S_o = GS_i \tag{3-26}$$

Then for the ideal case of Fig. (3-5) we have:

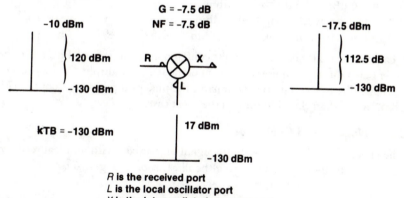

R is the received port
L is the local oscillator port
X is the intermediate frequency port

Fig. 3-5. Ideal signal case of mixer operation where the bandwidth is 25 kHz.

N_o = kTBFG = –130 dBm + 7.5 dB – 7.5 dB = –130 dBm
S_o = S_i G = –10 dBm – 7.5 dB = –17.5 dBm
Then

S_o/N_o = –17.5 dBm –(–130 dBm) = 112.5 dB

Note that the input signal to noise ratio is –10 dBm – (–130 dBm)=120 dB. This illustrates that kTB noise cannot be attenuated by a loss such as negative gain, and that the mixer output signal to noise ratio is degraded by the mixer noise figure.

3.5.2 Noisy Received Signal

Let the input received signal of Fig. (3-5) be corrupted by noise such that the previous signal of –10 dBm now has a noise floor of –90 dBm, all other conditions remaining the same. This case is shown in Fig. (3-6).

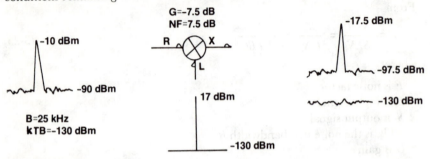

Fig. 3-6. Mixer output for the case of an ideal local oscillator signal and a non-ideal received signal.

From (3-25) and (3-26)

S_o = –10 dBm –7.5 dB = –17.5 dBm

N_o (ideal received signal) = –130 dBm

N_o (noise due to the signal) = –90 dBm + (–7.5 dB) = –97.5 dBm

It can be seen that there are two noise contributors. One being kTBGF and the other being due to the noise component of the signal, multiplied by gain. Since N_o (noise due to the signal) is dominant, the mixer output noise floor is –97.5 dBm, rather than the –130 dBm of the ideal case.

3.5.3 Noisy Local Oscillator

The effect of a noisy local oscillator signal on a mixer with an ideal received signal may be found by considering the illustration of Fig. (3-7).

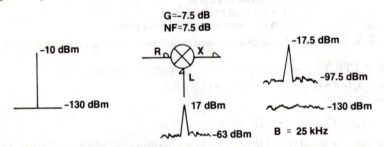

Fig. 3-7. An illustration of mixer behavior with a noisy local oscillator signal, and with an ideal received signal.

From (3-25) and (3-26)

S_o = -10 dBm - 7.5 dB = -17.5 dBm

N_o (noise due to signal) = -130 dBm + 7.5 dB - 7.5 dB = -130 dBm

Because the LO signal is not noise free, the local oscillator signal may be considered to be the sum of many LO signals. Therefore, conversion of the received signal would be expected over the whole of the LO signal. The result is a mixer output whose spectral profile is an emulation of the LO signal. The result of this is a raising of N_o to the S/N of the LO, which in this case is 80 dB. Once again there exist two noise sources, at the output port, of -130 dBm (ideal) and -97.5 dB, due to the LO noise (the latter dominates). A second consequence is the conversion of unwanted signals near the desired received signal by the broad noisy LO signal.

The effective noise figure of the mixer in this example becomes (using equation (3-25)):

$F = N_o/GN_i$
   = -97.5 dBm -(-7.5 dBm)-(-130 dBm)
   = 40 dB

Note, as the level of the input signal S_i diminishes, the mixer output signal noise floor also moves down dB for dB and eventually the LO noise floor no longer affects the output signal, which becomes dominated by kTBFG. Also, the effective noise figure of the mixer approaches that of the mixer with ideal signals. Thus low level operation of a mixer is not affected by LO noise. This bold statement must be qualified by requiring the LO signal be band limited to exclude any frequencies at the IF frequency. This prevents LO noise from entering the IF output of the mixer through the mixer's L to X port leakage.

3.6 NOISE MEASUREMENT TECHNIQUES

There are several methods of noise figure measurement. All are based on the injection of a known amount of noise into a receiver and observing the receiver output behavior. The source of this noise is almost universally white and is thermal noise. Thermal noise is a result of random electron motion. This noise exhibits a uniform energy distribution in the frequency domain and therefore has a constant power density with a normal or Gaussian level distribution.

A noise source should be capable of producing a significant noise power exceeding that of the device being measured. There are several popular noise sources, including: forward biased semiconductor diodes, temperature limited diodes, gas discharge tubes, and hot/cold resistive sources. The sources most often used are the diode and gas discharge types. They are useful to several hundreds of MHz and tens of GHz, respectively.

The temperature limited diode noise power may be predicted from:

$$P_n = kTB + (eIBR)/2 \tag{3-27}$$

where

kTB is termination resistance noise power for $T = 290°K$
e is the charge of an electron $(1.59 \cdot 10^{-19}$ coulomb)
I is the average diode current in amperes
B is the effective noise bandwidth in Hz

The noise power of the gas discharge source is:

$$P_n = kTB + k\,(T_1 - T)\,B_n \tag{3-28}$$

where

B_n is the overall source bandwidth
T_1 is the effective temperature of the gas discharge in degrees Kelvin

All other terms have been previously defined.

Noise factor may be computed from:

$$F = \frac{\dfrac{T_n}{290°\ Kelvin} - 1}{Y-1} = \frac{(constant)}{Y-1} \tag{3-29}$$

where

T_n is the effective temperature of the noise source
Y is the ratio of the output power of the device under test (with the noise source connected to the input) to the output of the device, with the input properly terminated at 290°Kelvin.

The only variable in the above equation is Y. Since the noise source is terminated with a load equal to that of the device under test,

$$Y = \frac{P_n\ on}{P_n\ off} \tag{3-30}$$

and

$$Y-1 = \frac{P_n\ on}{P_n\ off} - 1 = \frac{P_n\ on - P_n\ off}{P_n\ off} \tag{3-31}$$

From this, noise factor may be measured by measuring Y-1 at the test unit's output. The measurement of noise power must be made using a true rms meter, such as one using a bolometer or barretter detector.

Automatic noise figure measurement is made by gating the noise source on and off, and measuring the noise power at the output of the device under test with a meter calibrated directly in noise figure. The implementation of an automated measurement system is shown in Fig. (3-8).

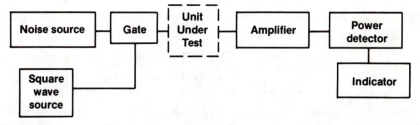

Fig. 3-8. Automatic noise figure measurement system.

3.6.1 Y Factor Method

One of the most accurate methods of noise figure measurement is the Y factor methods of Fig. (3-9).

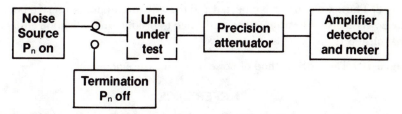

Fig. 3-9. Y factor method of noise figure measurement.

The attenuator is adjusted to provide the same output for P_n on and P_n off. The differential value of attenuation required to do this is Y. Then

$$F = 10 \log ((\ T_n / 290°\ Kelvin\ ) - 1\ ) - 10 \log (\ Y - 1\ ) \qquad (3\text{-}32)$$

where Y is a ratio.

This method provides accuracies of .1 to .2 dB.

3.6.2 3 dB Method

The 3 dB method determines noise figure by measuring the output power of the unit under test at properly terminated conditions, as shown in Fig. (3-10).

$$F = P_n - A_x \qquad (3\text{-}33)$$

The noise source is connected and a 3 dB attenuator is inserted before the power meter. The noise attenuator A_x is adjusted to provide a reading equal to that obtained earlier. The noise power into the receiver, or $P_n - A_x$, is equal to kTBF.

Measurement equipment is generally of fixed frequency and requires conversion to measure noise figure at other frequencies. Since the noise sources are broad band, a mixer, when tested for noise figure, will produce erroneous results because the image noise power will add to the desired noise power and provide double sideband noise figures. This may be corrected for by adding 3 dB to the number obtained, which results in the single sideband value.

When using mixers in the test setup, it is vital that the local oscillators have spectral purity or the results obtained will be erroneous.

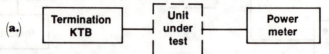

Step 1. The output power is measured with the unit under test properly terminated.

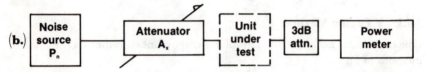

Step 2. The noise source is connected and the 3 dB attenuator is inserted before the power meter. The attenuator is adjusted to produce an indication equal to that of step 1.

Fig. 3-10. The 3 dB method of noise figure measurement.

REFERENCE

[1] *World Distribution and Characteristics of Atmospheric Radio Noise,* CCIR Report 322, 10th Plenary Assembly, Geneva, Switzerland, 1963.

4

THE RECEIVER

While there are several forms of receiving systems, none are more widely used than the superheterodyne. This chapter deals with the various forms of the superheterodyne, the constraints and considerations of the system's critical functional blocks, and the characteristics by which the system's performance is measured.

4.1 THE SUPERHETERODYNE

Practically all receivers in use today are of the superheterodyne configuration. As the name implies, a heterodyne or mixing process is involved. In this process the input signal is mixed with the output of an oscillator to produce an output frequency.

Mathematically this process is described by:

$$IF = | f_r \pm f_1 | \qquad (4\text{-}1)$$

where

IF is the output or intermediate frequency

f_r is the received frequency

f_1 is the local oscillator frequency

The process results in principal output terms at two IF frequencies. These terms are the absolute value of the sum (+) and the difference (−) of f_r and f_1. In practice only one is used and the desired output frequency is selected by a filter. The other is called an image frequency.

The major advantage of the superheterodyne configuration is that the bulk of the receiver's gain is placed at a single intermediate frequency (IF), rather than at a variable frequency as used in early tuned radio frequency receiving systems. The output of the IF amplifier is followed by detection and amplification.

4.1.1 Configurations

The superheterodyne has two basic forms which are:
down converter and
up converter

Each of these configurations may involve one to three or more conversions. As will be shown, the down converter can be accomplished usually in one conversion, but up converters utilize at least two in the usual case.

The basic superheterodyne configuration is shown in Fig. (4-1).

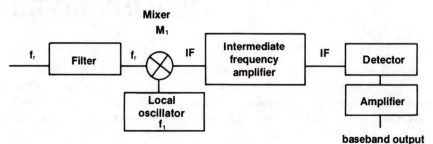

Fig. 4-1 The basic superheterodyne configuration.

The input signal f_r is selected by a filter and fed to a mixer (M1) where it is mixed with the local oscillator signal (f_l), to produce the intermediate frequency (IF). The output of the IF amplifier is detected to recover the modulation or baseband signal which is amplified and furnished as an output.

4.1.1.1 *Down Converter*

The down converter was the original realization of the superheterodyne receiving configuration, and it is almost exclusively used in home entertainment products, as well as in some high performance designs. The limit to its usefulness is preselection difficulties in wide band systems, because the preselector must be tunable. Tunability of the preselector also becomes a problem at the higher frequencies.

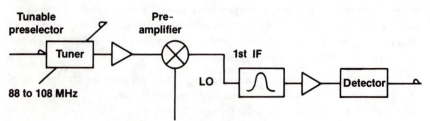

Fig. 4-2. Basic down converter configuration.

An example of a down converter design is shown in Fig. (4-2). This represents a typical FM home entertainment receiver. The input tuning range is 88 to 108 MHz and the intermediate frequency is 10.7 MHz. The advantage of down conversion is low cost and simplicity. However, it is not always the best choice.

4.1.1.2   Up Converter

By definition up conversion is the conversion of a frequency or band of frequencies to a higher intermediate frequency. This process results when the intermediate frequency is greater than the received frequency. As an example, a 2 to 30 MHz receiver using an IF of 120 MHz represents an up conversion configuration.

Up conversion is the trend in modern wide band high-performance systems. Hardware and component breakthroughs make it practical well into the GHz range. An example of up conversion is shown in Fig. (4-3).

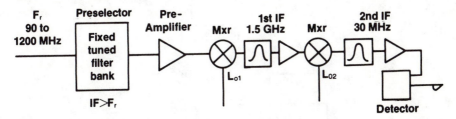

Fig. 4-3. An illustration of an up converter receiving system.

4.1.1.3   The Wadley Drift Canceling Local Oscillator System

Consider the up conversion receiver of Fig. (4-4). An input signal in the range of 30 to 400 MHz is converted by mixer M1 to a first IF of 1300 MHz, by a variable synthesized first local oscillator F_{lo}, whose frequency covers the range of 1330 to 1700 MHz. A second conversion to the second intermediate frequency of 110 MHz is provided by mixer M2 and a second local oscillator F_{lo2}, operating at a frequency of 1190 MHz. The first local oscillator being synthesized is based on a temperature compensated crystal oscillator (TCXO), whose accuracy is 1 ppm (part per million). The frequency error of F_{lo1} can then be expected to be as high as $1700 \cdot 10^{6} \cdot 1 \cdot 10^{-6}$ or 1700 Hz. Assuming that the cost is a major consideration, it would then be desirable not to unduly complicate the design by deriving the second local oscillator from the TCXO with a second phase-locked loop. Using a crude oscillator, for the second local oscillator, a drift of 25 ppm is expected. This results in a maximum frequency error of the second local oscillator of $25 \cdot 1190 = 29{,}750$ Hz. The total maximum error of the receiver is $29{,}750 + 1{,}700 = 31{,}450$ Hz. Errors of such magnitude can be an appreciable part of the receiver bandwidth. Should the final IF bandwidth be

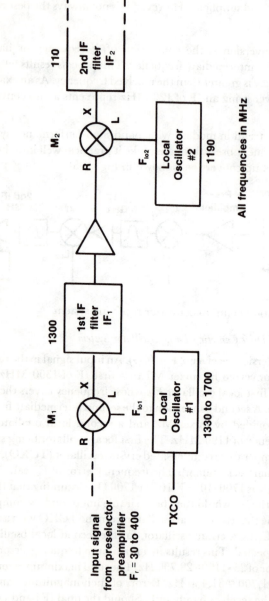

Fig. 4-4. A basic up conversion receiver showing the first two conversions.

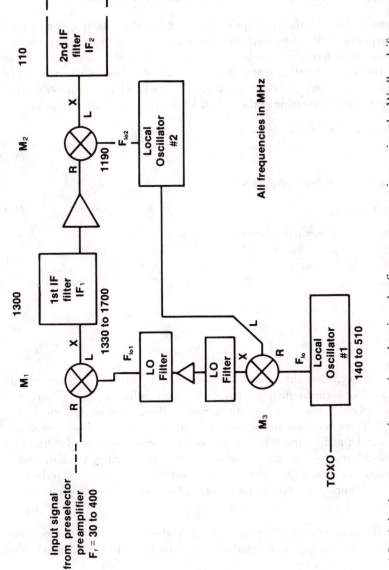

Fig. 4-5. A basic up conversion receiver showing the first two conversions using the Wadley drift canceling system.

25 kHz and the receiver tuned to a particular frequency; a desired signal, even though present in the RF spectrum, would not be received. One solution is to phase lock the second local oscillator to the TCXO. This reduces the second local oscillator error to 1.19 kHz, but at considerable expense.

A second solution is the Wadley drift canceling local oscillator system. This is shown in Fig. (4-5). The system is the same as that of Fig. (4-4), except the first local oscillator frequency has been reduced to 140 to 510 MHz and is mixed with the second local oscillator frequency of 1190 MHz in mixer M_3. The output of this mixer, as before, is the sum of the two inputs or 1330 to 1700 MHz. The drift canceling of the second local oscillator results as follows:

The first local oscillator frequency is:

$$F_{lo1} = F_{lo2} + F_{lo} \qquad (4\text{-}2)$$

The first intermediate frequency is

$$IF_1 = F_{lo1} - F_r \qquad (4\text{-}3)$$

The second intermediate frequency is

$$IF_2 = IF_1 - F_{lo2} \qquad (4\text{-}4)$$

Combining,

$$IF_2 = (F_{lo1} - F_r) - F_{lo2}$$
$$IF_2 = F_{lo2} + F_{lo} - F_r - F_{lo2} \qquad (4\text{-}5)$$
$$IF_2 = F_{lo} - F_r$$

It can be seen that the second intermediate frequency (IF_2) is completely independent of the frequency of the second local oscillator (IF_{lo2}). Clearly any frequency errors resulting from the second local oscillator are cancelled completely. There is a limit to the allowable error of the second local oscillator, which is defined by the first intermediate frequency filter bandwidth. This bandwidth must accommodate: its own drift, the desired signals bandwidth, the frequency error of the incoming signal and the errors of the first and second local oscillators. This is seldom a problem, but it should be considered in the design.

Note, the tuning range of the two systems synthesized local oscillators, differ. The conventional approach tuned over a ratio of 1700/1330 = 1.278, while for the Wadley system this ratio is 510/140 = 3.64. While the 1.278 ratio can be accommodated by a single voltage tuned oscillator (VTO), the 3.64 tuning ratio requires switching between several VTOs. This results in additional complexity.

The number of VTOs can be reduced by allowing the Wadley mixer M_3 to operate in both the sum and difference modes. The design modifications are as follows:

Make the second intermediate frequency equal to the receiver's tuning range divided by four, or in this example (400–30)/4 = 92.5 MHz.

The first mixer is allowed to operate in both the sum mode for the lower half of the receiver tuning range and in the difference mode for the upper half of that range.

Example:
    30 to 215 MHz
    Mixer M_1 operates in the sum mode
    215 to 400 MHz
    Mixer M_1 operates in the difference mode.

The first local oscillator frequencies are:

Band	F_r (MHz)	F_{lo1} (MHz)
1	30 to 215	1270 to 1085
2	215 to 400	1515 to 1700

The second local oscillator frequency is the sum of the first and second intermediate frequencies.

Example: 1300 + 92.5 = 1392.5 MHz = F_{lo2}

The VTO frequencies are:

Band 1 $F_{lo2} - F_{lo1}$
        or 1392.5 – 1270 to 1085 = 122.5 to 307.5 MHz

Band 2 $F_{lo1} - F_{lo2}$
        1515 to 1700 – 1392.5 = 122.5 to 307.5 MHz

The revised system is shown in Fig. (4-6). Note a trade off between the number of VTOs (reduced tuning range) and filters was made.

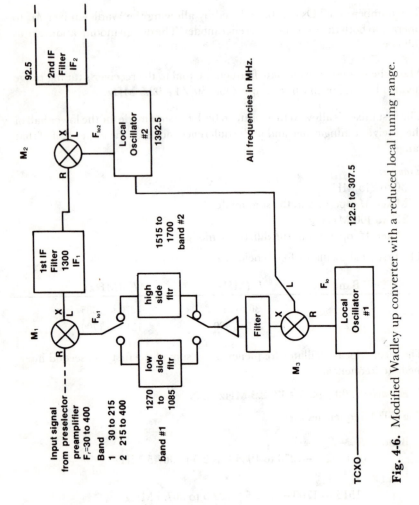

Fig. 4-6. Modified Wadley up converter with a reduced local tuning range.

4.2 DIRECT CONVERSION RECEIVERS

This is a special class of receiver which employs superhetrodyne principles but utilizes a zero IF. The local oscillator frequency is made equal to that of the signal.

Since IF = $F_r \pm F_{lo}$ for a superheterodyne receiver and for a direct conversion receiver IF = 0, then:

$$0 = F_r - F_{lo} \tag{4-6}$$

or

$$F_r = F_{lo} \tag{4-7}$$

where

IF is the intermediate frequency
F_r is the signal frequency
F_{lo} is the local oscillator frequency

The implementation of such a receiver is shown in Fig. (4-7).

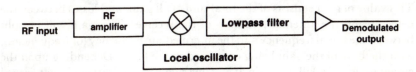

Fig. 4-7. Basic direct conversion receiver.

The signal from the RF amplifier is fed to a mixer (or product detector) whose local oscillator frequency is made equal to that of the incoming signal. The output of the mixer is the demodulated RF input signal and does not require any other detection. A simple lowpass filter of the RC variety provides the required IF selectivity. The cutoff frequency is slightly above the highest demodulated frequency of interest. This baseband signal is amplified and provided to the output terminals.

Such low cost simple receivers are generally not used because of several serious limitations:

The local oscillator stability requirements restrict its usefulness to frequencies below 20 MHz.

They can demodulate forms of AM signals only such as on/off keying, SSB, and AM DSB. The latter can only be demodulated if care in tuning is exercised. The detection of FM and PM is not feasible using direct conversion techniques.

Direct conversion must not be used without RF amplification in order to reduce local oscillator levels present at the receiver's antenna terminals to tolerable limits.

4.3   THE RECEIVER RF AND IF GAIN BUDGET

The receiver detector serves as a convenient point of reference for the determination of the gain distribution of the receiver. After selecting the input signal requirements of the detector, the combined RF and IF gain needed to amplify the input signal at the receiver antenna terminals may be computed.

For diode detection, an input power level of −10 to 0 dBm is generally adequate. For high sensitivity receivers where IF gain becomes critically high, −10 dBm is a good choice.

For example, if the minimum signal power is −98 dBm, and a detector input power level of − 10 dBm, allowing a 10 dB margin, the required combined RF and IF gain must be:

$$(-98 - (-10) - 10) = 98 \text{ dB}$$

This value of gain consists of the total sum of all gains and losses between the detector and the antenna terminals of the receiver. This gain will be split between the signal frequency and the intermediate frequency (or frequencies), with the bulk of the gain being contained in the latter. Depending upon the circuit designer's skill, the gain at any one single frequency should not exceed 100 dB. Where greater gain is required, it is preferred to split it between several amplifiers at different frequencies. If the required gain at any signal IF is found to be unmanageable, an extra conversion to another frequency is usually a good solution.

4.4   PRESELECTOR REQUIREMENTS

The preselector must reduce the receiver's response to the image frequency, which is twice the intermediate frequency away from the desired frequency. The preselector must also reduce the radiation of the receiver's local oscillator out of the receiving antenna (which is only the intermediate frequency away from the tuned receiver frequency). Additionally, it must reduce the receiver's response to the intermediate frequency and its subharmonics, as well as providing sufficient rejection to spurious responses. Such responses are caused by the mixing process, while not attenuating the desired input frequency. The specific needs of preselection differ between up and down conversions, as explained in the following sections.

4.4.1   Preselection Requirements for Down Converters

Because a down converter uses as the first IF a frequency lower than that being received, wide band receivers using a single conversion usually require a tunable preselector. Where multiple down conversions are used, such as for UHF, VHF or microwave bands, or for narrow reception band applications, the design becomes a border line case. In such cases a complete performance analysis must be made. This includes the response of the receiver to unwanted signals such as spurious response images and IF rejection, *et cetera*.

Tunable preselection is usually required in down conversion applications. This is a result of the fact that the LO and image frequencies are usually in band. Because the IF must be out of band, IF rejection is not a serious problem. The subharmonics of the received signal, mixing with the fundamental of the LO, is a serious problem.

Consider a receiver operating in the 30 to 200 range and down conversion is selected, although this may not be the best solution. The first IF must be less than 30 to be out of band. A popular frequency where IF filters are readily available is 21.4 MHz. Using this frequency, the LO frequency range is computed using high side injection (LO above signal).

$$LO = F_r + IF \tag{4-8}$$

where

F_r is the received frequency

then

$$LO = 30 + 21.4 \text{ to } 200 + 21.4$$
$$= 51.4 \text{ to } 221.4 \text{ MHz}$$

It can be seen that the LO is in band over a considerable portion of the receiver tuning range. It must not be allowed to radiate out through the receiver antenna causing interference to other receivers and perhaps, in some cases, allowing direction finders to locate the receiver.

The image frequencies (in this case) are two times (the IF) above the receiver frequency, and may be found from:

$$F_{image} = F_r + 2IF \tag{4-9}$$

or

$$F_{image} = LO + IF \tag{4-10}$$

from (4-10)

$$F_{image} = 51.4 + 21.4 \text{ to } 221.4 + 21.4 \text{ MHz}$$

which is largely in band.

Low order subharmonics of F_r, mixing with the LO, will cause serious spurious products and must be attenuated by the preselector. This applies also to the case where the IF subharmonics fall into the preselector passband. The most serious case results at the 1/2 IF and the IF itself.

The summary of frequencies which must be at the preselector's ultimate attenuation zone are (where n is an integer):

Intermediate Frequency / n
Received Frequency / n
Local Oscillator Frequency
Image Frequency

It can be seen that the IF determines the width of the preselector filter's ultimate attenuation bandwidth, as does the received frequency.

Considering the IF alone, the preselection type is a function of frequency. Low IFs are best filtered by tunable means. Higher frequencies are filterable by either fixed tuned or tunable filters, except at high RF input signals, where the fixed tuned filter is best.

It is most economical to utilize tuned preselection which tracks the receiver tuning. Tuned preselection is limited in tuning range to less than 3 to 1; therefore several tuned preselectors are required. The number of preselection bands would be:

$$F_{r\,min}\,K^n = F_{r\,max} \tag{4-11}$$

then

$30\,K^n = 200$
$K = \sqrt[n]{6.666}$, where $K < 3$
$n = 2$ gives $K = 2.58$

where

K is the tuning range
n is the number of filters

Therefore two to three switched tunable preselection filters are required. While some designs use mechanical tuning of capacitance or inductance, the trend, in modern equipment, is to use varactor or digital tuned electronic tuning.

4.4.2 Preselection Requirements for Up Converters

Because of a high intermediate frequency, the up converter has the advantage of simpler preselection requirements. The first constraint is the $m = 1$ and $n = 2$ spurious responses which for a doubly balanced mixer is typically –67 to –83 dBc. This 1 by 2 spur is the desired receiver frequency divided by 2; mixing with the fundamental of the local oscillator. Should –67 to –83 dBc be considered satisfactory the designer must consider the 1 by 3 case, where the response is typically –49 to 90 dBc, depending on the mixer used and the local oscillator power. In either case, the preselector will have to make up any additive attenuation plus margin. If the 1 by 2 case is the limiting case, the preselector filter bandwidth must be less than 2 to 1 (or less than 3 to 1 for the 1 by 3 case). A simple fixed tuned filter will usually suffice for upconversion receiver designs of high performance. A value of 1.5 to 1 is generally sufficient for $n = 4$, 0.1 dB ripple Chebyshev preselector filtering. If a more detailed analysis is desired; the attenuation at the high end of the band to the response at 1/2 of this value can be computed and added to the value given in a mixer spur table. A similar procedure is used if the 1 by 3 case is the limit. When the 1 by 2 case is the limit, the 1 by 3 case is automatically satisfied.

For example, consider a receiving system which covers a tuning range of 90 to 450 MHz and the one by two case is the limit. Here, the number of fixed tuned preselector filters is computed from

$$(f_{min})\ K^n = f_{max} \text{ where } K \approx 1.5 \tag{4-12}$$

and

n is the integer number of filters

Then

$$K = \sqrt[n]{f_{max}/f_{min}} \tag{4-13}$$

In this example,

$$K = \sqrt[n]{450/90} = \sqrt[n]{5}$$

from which $K \cong 1.49$ if $n = 4$

Therefore four filters are required and are defined as follows:

Band	MHz.
1	90 to 1.49 · 90 = 90 to 134
2	134 to 1.49 · 134 = 134 to 200
3	200 to 1.49 · 200 = 200 to 298
4	298 to 450

Local oscillator radiation supression by the preselector for up conversion is generally limited by the ultimate attenuation of the preselector. As in the case of the first IF frequency, being reasonably higher than the highest frequency to be received. The attenuation of the local oscillator frequency by the preselector must be examined on an individual basis to determine the effectiveness of the preselector.

The IF rejection of the preselector is usually limited by the highest preselector band filter. The filter characteristic must be examined at the IF to determine its attenuation, which should be at the filter's ultimate value. It is not uncommon to add more sections to the filter to achieve this goal.

Image rejection is usually not a problem because of the high first IF. To verify this, the designer should compute the image frequency band and compare it to the preselector filter characteristics. The preselector should present its ultimate attenuation to the image frequencies.

The spurious response of the receiver should be examined on a band by band basis. The fact that the LO and IF are high results and that most of the troublesome spurs are out of band, they therefore suffer the ultimate attenuation of the preselector.

Most preselectors offer only about 50 or 60 dB isolation between bands because of switching imperfection which effectively limit the ultimate attenuation of the filters to that value. It is usually found necessary to add an additional lowpass filter, to secure the desired level of performance.

An example of an up converter is shown in Fig. (4-8).

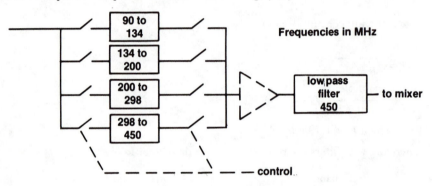

Fig. 4-8. A typical up converter preselector configuration.

4.5 THE NEED FOR AN OUT OF BAND INTERMEDIATE FREQUENCY

The conversion process where the signal frequency is changed to the intermediate frequency is described by:

$$F_{if} = \left| F_r \pm F_l \right| \tag{4-14}$$

where

F_{if} is the intermediate frequency and is greater than 0
F_r is the receiver frequency
F_l is the local oscillator frequency

Note: The associated sign may be either + or − but not both for any single frequency. If the received signal was equal to that of the intermediate frequency amplifier then:

$$F_{if} = F_r$$

and

$$F_l = \left| F_{ij} \pm F_{ij} \right| = 0, 2 F_{if}$$

It is not practical to let $F_l = 0$ in any tunable receiving system. The reason being, at some other value of F, a non-zero value of F_l results, and the required local oscillator tuning ratio becomes:

$R = F_l / 0$, which is infinity and is not realizable

Further, a mixer operated with $F_l = 0$ is no longer a mixer but a switch, which is on all of the time and may as well be deleted. The receiver then reduces to a tuned amplifier, followed by a detector, which is suitable for fixed frequency operation. The exception to this discussion is the direct conversion receiver which uses a 0 IF value described elsewhere.

Considering the case where $F_{lo} = 2F_{if}$, there are two signal outputs from the mixer at F_{if}; one being the converted signal and the other the mixer leakage signal, which is unconverted. The levels of these signals for a good quality double balanced mixer are −6 dB and −20 to 25 dB, respectively. Because of slight frequency inaccuracies of the transmitted frequency and the receiver local oscillator frequency, F_r will not be exactly equal to F_{if}. Therefore a very serious heterodyne results.

A ratio of 0.5 between the received frequency and the local oscillator should be avoided, because of the very serious multiplicity of spurious products created with this ratio (see section 4.27). Therefore, it is good design practice to avoid having an intermediate frequency which is included within the tuning range of the receiver.

4.6   THE LOCAL OSCILLATOR FREQUENCY

The local oscillator frequency F_l, is dependent upon the input frequency to the mixer. The intermediate frequency IF as defined by:

$$F_l = \left| IF \pm F_r \right| \tag{4-15}$$

The sign is dependent upon the choice of the sum or difference mixing. The difference mode is always employed for down converters, with the local oscillator being either above or below the received signal. In up conversion, either sum or difference mixing may be employed.

It is mandatory that the local oscillator (LO) low order harmonics must not be equal to the first IF. If equal, serious beats or birdies will result and AGC take-over may desensitize the receiver at these points.

4.7   AUTOMATIC GAIN CONTROL

Where the modulated signal is represented by variations in signal magnitude such as: amplitude modulated carriers, double sideband suppressed carriers, and single sideband suppressed carrier signals, the amplitude of the signal components must be retained and linear amplification must be utilized. Failure to maintain the amplitude variations through the receiver will result in distortion or total loss of the modulation.

Because the signal must be amplified 100 dB or more, and signal strength may vary equally, it is not possible for an amplifier to cope with the situation for all cases. To illustrate, assume an amplifier was designed to process a signal of –93 dBm and has a 100 dB gain in order to provide the required output level of 7 dBm or 1/2 volt. The amplifier power supply is 12 volts dc. Should a strong signal of 0 dBm be encountered, the output would have to be 0 dBm +100 dB or 100 dBm, which is in excess of 10,000 volts. If it were possible to accomplish such a feat, an output level control would have to be adjusted every time a signal level changed, due to fading or a different signal with a different signal level.

In the usual case the output of an amplifier is limited by its power supply voltage. An input signal level which causes the last stage of the amplifier to become non-linear, because it can no longer duplicate the input signal variations, results in distortion caused by flattening out the signal peaks.

As the input signal level is increased the distortion increases until all signal amplitude variations are totally lost, and the intelligence to be transmitted is no longer recoverable. This effect is often called compression blocking or limiting, and is undesirable for amplitude modulated signals.

A technique which can overcome this problem does so by automatically adjusting the gain of the amplifier directly proportional to that of the signal strength. This technique is called Automatic Gain Control or AGC. The gain of the receiver is controlled by the signal strength at the receiver output. Basically, the output of the detector is compared to a reference. Any differential is amplified and fed back to the previous stages as a control signal to vary the gain of the receiver such that the output is constant over the range of expected signal intensity. This is illustrated in Fig. (4-9).

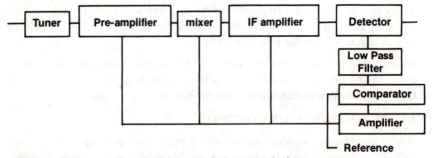

Fig. 4-9. An illustration of automatic gain control of the receiver.

4.7.1 Determining the AGC Control Range

To determine the dynamic range of the AGC system it is necessary to establish:

the minimum signal power level (S_{min}) and
the maximum signal power level (S_{max})

The ACG control (Δ_{AGC}) is therefore

$$S_{max} - S_{min} = \Delta_{AGC} \tag{4-16}$$

4.7.2 Attenuation AGC

The use of an attenuator for AGC overcomes the problem of dynamic range variation due to operating point shift of the devices used in the amplifier. This shift of operating point from high gain to low gain in either the forward or reverse AGC modes reduces the output swing of the amplifier for strong signal cases. While this is of no consequence in most cases, it is of vital concern to the designer striving for larger values of instantaneous dynamic range. For the latter case, an amplifier of fixed gain is used with one or more voltage variable attenuators. The location of such attenuators requires careful planning. The basic concept of attenuation AGC is shown in Fig. (4-10).

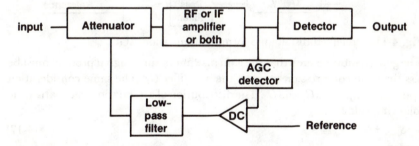

Fig. 4-10. Basic concept of AGC using attenuation provided by voltage variable attenuators.

The attenuator is placed ahead of the RF-IF amplifier to prevent the input signal from ever saturating the amplifier. The AGC detector provides an output voltage proportional to signal strength, which is fed to a DC differential amplifier. The second input of the amplifier is provided with a reference voltage which establishes the output of the detector. The dc amplifier's output is lowpass filtered with a time constant long enough to prevent instability of the loop and destruction of the low modulation frequencies of the signal, which the AGC voltage must not be allowed to follow. This filtered voltage is the AGC voltage, whose magnitude is related to signal strength. It is applied to the attenuator to control its attenuation, providing attenuation directly proportional to signal strength. The result is the amplifier input voltage is nearly constant with a flatness depending upon the AGC loop gain.

While Fig. (4-10) illustrates the concept, it is not the most desirable implementation. Since the signal input power to the amplifier is held constant, and the noise generated in the amplifier is constant, the output signal to noise ratio is also constant, regardless of signal strength. This is unacceptable except for low quality links.

By distributing the attenuation within the amplifier in two or more blocks, the overall noise figure, which with the configuration of Fig. (4-10) increased directly with the signal power dB for dBm, can be buffered by gain and held nearly fixed. The output signal to noise ratio can be made to increase with input signal power. Fig. (4-11) illustrates this preferred arrangement.

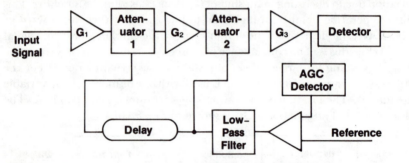

Fig. 4-11. Preferred distributed form of attenuation AGC.

The gain G_1 must be such that the gain maximum times signal product must be less than the compression point of the amplifier G_1. The same consideration applies to amplifiers G_2 and G_3. The distribution of attenuation must satisfy the following rules:

$$S_{i\,max} + G_i = S_{o1} \leqq S_{o1}' \tag{4-17}$$

$$S_{o1} + \alpha_1 + G_2 = S_{o2} \leqq S_{o2}' \tag{4-18}$$

$$S_{o2} + \alpha_2 + G_3 = S_{o3} \leqq S_{o3}' \tag{4-19}$$

or generally

$$S_{oN} + \alpha_n + G_{n+1} = S_{o\,n+1} \leqq S_{o\,n+1}' \tag{4-20}$$

where

    S_i is the maximum input power to the first stage (dBm)
    G_1 is the gain of the input amplifier (dB)
    S_{o1} is the output signal power from the first stage (dBm)
    S_{o1}' is the −1 dB compression point of the first stage
    G_2 is the gain of the second amplifier (dB)
    α_1 is the maximum attenuation of the first attenuator (dB)
    n is the nth stage, *et cetera*

To result in the lowest noise output or maximum quieting with increasing signal strength, the AGC should be applied beginning with the attenuator closest to the output. The sequence of attenuation call up should be α_n ...... α_3, α_2, α_1. This is accomplished by introducing delay in the AGC buss to the earlier attenuators. Generally two to three attenuation sections are adequate for most requirements. This procedure maintains the lowest overall noise figure with increasing signal power and it prevents the compression of any stage.

4.7.3 Fast Attack/Slow Decay AGC

The fast attack/slow decay AGC from involves circuit modifications, which cause the AGC voltage to build up rapidly with the application of a signal and down slowly upon removal of the signal. This form of AGC is generally used with signal sideband suppressed carrier signals. In special cases it is used with conventional AM.

There are many ways of implementing this form of control. Some of the more popular forms are shown in Fig. (4-12). Essentially, the time constant of the normal lowpass AGC filter is replaced with a dual time constant lowpass filter, where the charge time (t_r) is less than the discharge time (t_f).

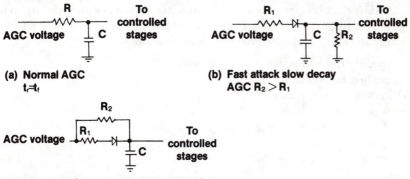

(a) Normal AGC
$t_r = t_f$

(b) Fast attack slow decay
AGC $R_2 > R_1$

(c) Fast attack/slow decay AGC $R_2 > R_1$ >

Fig. 4-12 AGC filter forms: (a) normal AGC low pass filter used in average AGC system; (b) and (c) fast attack/slow decay AGC filter forms.

Since AGC utilizes negative feedback, it is important to choose the attack time constant so that the desired baseband signals do not appear on the AGC buss. If they do, they will be removed or mutilated at the receiver's output.

Because of differing attack and decay time constants, the system tends toward peak detection of the signal and is no longer an averaging AGC (which refers to the carrier and not the modulation peaks). Fast attack/slow decay AGC establishes its reference to the signal peaks; thus, this form of AGC is amplitude modulation dependent. Suppressed carrier systems have little or no carrier but contain sideband energy whenever modulation is present. At all other times, for all practical purposes, there is no carrier. Thus, fast attack / slow decay AGC is mandatory for such applications. It is optional for all other forms.

4.8 SENSITIVITY

Sensitivity is a measure of how weak a signal can be and still be received satisfactorily. The limit to the sensitivity of a receiver is noise, both external and internal. Since external or site noise is beyond the control of the designer, only internal receiver noise is considered in the sensitivity specification. The usual forms of specifications for this parameter are given as:

n_1 microvolts input for a n_2 (dB) S/N
n_1 microvolts input for a n_2 (dB) $(S + N)/N$
n_1 microvolts input for a n_2 (dB) SINAD.

In some cases the input signal strength may be given in dBm rather than microvolts.

Where

S/N is the output signal to noise power ratio.

$(S+N)/N$ is the output signal plus noise to noise ratio.

SINAD is an abbreviated form of the output signal plus noise and distortion to noise and distortion power ratio. The unit dBm is the signal power level relative to 1 milliwatt. It is also necessary to specify the following:

receiver input impedance (usually 50 Ω)
modulating frequency
amount of modulation
type of modulation

S/N and $(S + N)/N$ are related by:

$$\frac{S+N}{N} = \frac{S}{N} + 1 \quad \textit{(power ratios)} \tag{4-21}$$

Table 4-1 tabulates this relationship for convenience, and a graph of this relationship is shown in Fig. (4-13).

4.8.1 Measuring Sensitivity Given S/N or (S + N)/N

A test setup suitable for the measurement of sensitivity is shown in Fig. (4-14).

The signal is tuned to the receiver frequency of interest and is modulated at a frequency and degree given in the test specifications. The modulation is turned on and off while the output of the receiver is measured on a true RMS voltmeter. The RF signal amplitude of the signal generator is adjusted until the required $(S + N)/N$ ratio is observed.

Note that the true RMS voltmeter measures $(S + N)/N$ which if required must be converted to S/N. Also it is mandatory that a true RMS meter is employed for this measurement, to give an accurate noise component reading.

Table 4-1.

S/N as Related to $(S+N)/N$

S/N, dB	S/N, ratio	$(S+N)/N$, ratio	$(S+N)/N$, dB
1	1.25	2.25	3.52
2	1.58	2.58	4.12
3	1.99	2.99	4.76
4	2.51	3.51	5.45
5	3.16	4.16	6.19
6	3.98	4.98	6.97
7	5.01	6.01	7.79
8	6.3	7.3	8.63
9	7.94	8.94	9.51
10	10	111	10.41
15	31.62	32.62	15.13
20	100	101	20.04

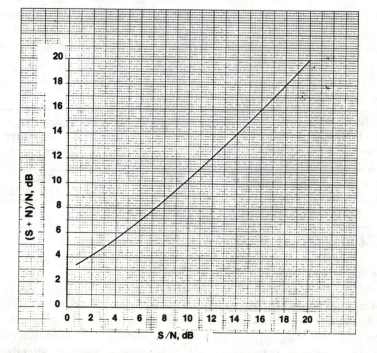

Fig. 4-13. Conversion of $(S+N)/N$ to S/N, dB and conversely.

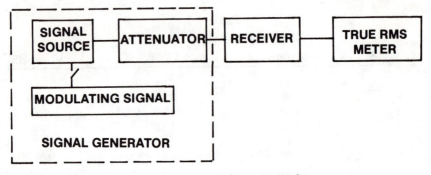

Fig. 4-14. Sensitivity test set up given S/N or $(S+N)/N$.

4.8.2   Measurement of Sensitivity Given SINAD

SINAD is a ratio of signal plus noise, distortion to noise and distortion in dB. This ratio may be given in numerical or dB form and is a power ratio. The only difference between the numerator and the denominator is that portion provided by the signal. Since this is a demodulated output measurement, it requires the removal of the fundamental of the modulating frequency; while retaining the noise and distortion components in the denominator. The test set up used for this measurement is shown in Fig. (4-15).

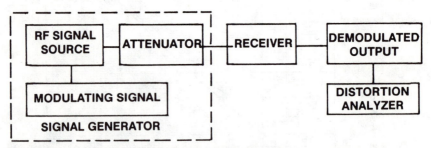

Fig. 4-15. Sensitivity test set up given SINAD requirements.

The signal generator is tuned to the receiver frequency of interest and is modulated at a frequency and degree given in the specificiations. The distortion analyzer is tuned to null out the modulation signal fundamental. Switching to the total input power position, the meter reads signal plus noise and distortion. Switching to the null position, the signal is nulled out, leaving the noise plus distortion components. The signal generator output attenuator is adjusted until the required SINAD is observed.

4.9 SIGNAL TO NOISE RATIOS FOR AMPLITUDE MODULATED DOUBLE SIDEBAND SYSTEMS

In section 2.1 it was shown that an amplitude modulated double sideband signal consists of a carrier and two sidebands. The power in each component of the wave which contributes to the total power is:

$$P_t = A^2 + \frac{A^2 m^2}{4} + \frac{A^2 m^2}{4}$$

(4-22)

where

P_t is the total power in the signal
A is the magnitude of the carrier
m is the modulation factor

or

$$P_t = A^2 + \frac{A^2 m^2}{2} = A^2 \left(1 + \frac{m^2}{2} \right)$$

(4-23)

The portion of the signal containing the modulation is:

$$P_{mod} = \frac{A^2 m^2}{2}$$

(4-24)

Let $P_c = A^2$ = the carrier power. The output signal to noise ratio becomes:

$$\frac{S}{N} = \frac{\dfrac{P_c m^2}{2}}{kTBF} = \frac{P_c m^2}{2kTBF}$$

(4-25)

Where

P_c is the carrier power
m is the modulation factor
1 = 100%, .5 = 50%, *et cetera*
k is Boltzmann's constant $(1.38 \cdot 10^{-23})$ joules / °K
T is 290° Kelvin
B is the post-detection effective noise bandwidth
K is degrees Kelvin
F is noise factor

In a more convenient dB notation,

$$\frac{S}{N} = P_c + m^2 - (3 + kT + B + NF)$$

(4-26)

Where

NF is noise figure = $10 \log F$
$10 \log 2 = 3$ dB

Example 4-1:

Given:

The signal strength is 3 μV across 50 Ω

The modulation percentage is 30

The post-detection effective noise bandwidth is 4 kHz (see section 3.2 for a discussion of ENB)

The signal plus noise to noise ratio is 10 dB.

Find the receiver's required noise figure.

The signal power is

$$P_c = \frac{(3 \cdot 10^{-6})^2}{50} = -127.44 \text{ dBw} = -99.744 \text{ dBm}$$

The value of S/N is found from

$$\frac{S+N}{N} = 10 \text{ dB} = 10 \text{ ratio}$$

$$S + N = 10 \text{ N}$$

or

$$S = 9N$$

and

$$S/N = 9$$

$$10 \log 9 = 9.54 \text{ dB}$$

A modulation percentage of 30 percent is a modulation factor of 0.3 = m

$$10 \log m^2 = -10.46 \text{ dB}$$

kTB is in dB

$$10 \log (1.38 \cdot 10^{-23} \cdot 290 \cdot 4 \cdot 10^3) = -168 \text{ dBw}$$
$$= -138 \text{ dBm.}$$

Rewriting (4-26) and solving for NF we have:

$$NF = -(S/N) + P_c + m^2 - kTB - 3$$

Substituting,

$$NF = -9.54 - 97.44 - 10.46 + 138 - 3$$

$$= 17.56 \text{ dB}$$

This represents the maximum receiver noise figure which will satisfy the requirements. Because of production variance the designer should allow a comfortable margin and design for a lesser value.

4.10   FM CARRIER TO NOISE RATIO

The carrier to noise ratio of a received signal is defined by:

$$\frac{FM\ carrier}{Noise} = \frac{P_c}{N_i} = \frac{P_c}{kTBF} \tag{4-27}$$

where

P_c is carrier power
kT is –144 dBm/kHz (used as a ratio)
B is the effective IF bandwidth (see section 3.2)
F is the receiver noise factor
N_i is the noise measured at the input of the limiter

In dB notation this ratio may be expressed by:

$$P_c - kT - B - NF \tag{4-28}$$

Example 4-2:

P_c = –60 dBm
kT = –144 dBm/kHz
B = 25 kHz = 14 dB
NF = 12 dB

The carrier to noise ratio is:

$$-60 - (-144) - 14 - 12 = 58\ dB$$

4.10.1   FM Output Signal to Noise Ratio Above Threshold

The FM outpt signal to noise ratio is related to the FM carrier to noise ratio by a term known as the modulation noise improvement ratio (MNI), as shown below:

$$(S/N)_{out} = (MNI)\ (P_c/N_i) \tag{4-29}$$

where

S is signal
N is noise
P_c is carrier power
N_i is noise measured at the input to the limiter

The modulation noise improvement factor is a term resulting from a trade-off between bandwidth and noise.

4.10.2   FM Noise Improvement Factor (MNI) Above Threshold

For carrier to noise ratios $> \approx 10$ dB, the modulation noise improvement factor is defined by:

$$MNI = \frac{3}{2} \left(\frac{\Delta F}{B_a} \right)^2 \left(\frac{B_{if}}{B_a} \right)$$

(4-30)

Where

B_a is the one-sided audio noise bandwidth
ΔF is the peak deviation of the carrier
B_{if} is the effective IF bandwidth

Example 4-3:

In a digital system, the IF bandwidth is 240 kHz and the data rate is 120 kbits/sec with a peak deviation of 300 kHz

$$MNI = \frac{3}{2} \left(\frac{300}{120} \right)^2 \left(\frac{240}{120} \right) = 18.75 = 12.73 \text{ dB}$$

The value of MNI can be improved by increasing deviation and IF bandwidth for a given upper modulation frequency. Reducing the modulation frequency for a given deviation and bandwidth improves MNI.

For a given deviation, the IF bandwidth must be changed to include all significant signal spectral terms. The necessary bandwidth is given by Carson's rule:

$$B_{if} = 2 (\Delta F + f_m) , \Delta F > 1$$

(4-31)

where

f_m is the modulation frequency

4.10.3   FM Signal to Noise Ratio Below Threshold

Where the carrier to noise ratio is less than 10 but more than 3, the output signal to noise ratio may be calculated by modifying the above threshold relationship as follows:

$$(S/N)_{out} = \frac{\dfrac{3}{2} \left(\dfrac{\Delta F}{B_a} \right)^2 \left(\dfrac{B_{if}}{B_a} \right) \left(\dfrac{P_c}{N_i} \right)}{1 + 0.9 \left(\dfrac{B_{if}}{B_a} \right)^2 \dfrac{(P_c/N_i) \, e^{-(P_c/N_i)}}{\left(1 - e^{-(P_c/N_i)} \right)^2}}$$

(4-32)

where all of the above terms were defined in the previous associated sections. See reference [1] for additional information.

4.11 PM OUTPUT SIGNAL TO NOISE RATIO ABOVE THRESHOLD

The PM output signal to noise ratio is related to the carrier to noise ratio by:

$$(S/N)_{out} = M^2 \ (B_{if}/(2 \ B_a)) \ (P_c/N_i) \tag{4-33}$$

where

S/N is the output signal to noise ratio
B_{if} is the IF noise bandwidth
B_a is the one-sided audio noise bandwidth
M is the peak phase deviation
P_c/N_i is the carrier to noise ratio in the IF noise bandwidth

This differs from the equation for FM by M^2 replacing $3(\Delta F/B_a)^2$ but is otherwise identical.

See reference [2] for additional information.

4.11.1 PM Output Signal to Noise Ratio below Threshold

Where the carrier to noise ratio is less than approximately 10, the output signal to noise ratio is defined by:

$$(S/N)_{out} = \frac{M^2 \left(\dfrac{B_{if}}{2B_a}\right) \left(\dfrac{P_c}{N_i}\right)}{1 + 0.9 \left(\dfrac{B_{if}}{B_a}\right)^2 \dfrac{(P_c/N_i) \ e^{-P_c/N_i}}{(1 - e^{-P_c/N_i})^2}} \tag{4-34}$$

All of the above terms have been defined in the preceding section.

Example 4-4:

Given:

$B_{if} = 21$ kHz
$B_a = 3$ kHz
$M = 5$

Compute the relationship between $(S/N)_{ou}$ and (P_c/N_i) over the range of 0 to 30 dB for (P_c/N_i).

The short computer program of Table 4-2 was executed and the print-out is shown in Table 4-3. The threshold of 10 dB for (P_c/N_i) is seen as the break point between the below and above threshold regions. Above this threshold the denominator of equation (4-34) is unity and may be omitted.

A second program with a plot routine is shown in Table 4-4 and the plot obtained for Example 4-4 is shown in Fig. (4-16).

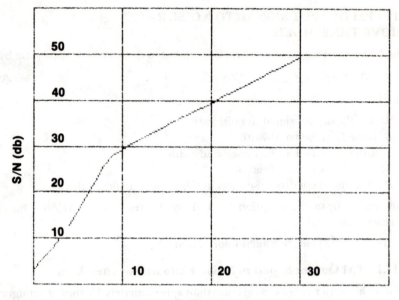

Fig. 4-16. Plot of Example 4-4 using the program of Table 4-4.

Table 4-2.

Computer Program for the Calculation of S/N as a Function of the Carrier to Noise Ratio and the Phase Modulation Index

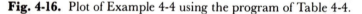

```
 10  SHORT T1,T2,D,C
 20  DISP "ENTER BIF,BA,M"
 30  INPUT B1,B2,M
 40  PRINT "  S1         S2        D
     C"
 50  FOR C1=1 TO 30 STEP 1
 60  C=10^(C1/10)
 70  S1=M^2*(B1/(B2*2))*C
 80  D=1+.9*(B1/B2)^2*(C*EXP(-C))
     /(1-EXP(-C))^2
 90  S2=S1/D
100  T1=10*LGT(S1)
110  T2=10*LGT(S2)
120  PRINT T1;TAB(8);T2;TAB(16);D
     ;TAB(24);C1
130  NEXT C1
140  END
```

Where

S1 is the signal to noise output ratio (dB) above threshold

S2 is the signal to noise output ratio (dB) below threshold

D is the factor relating S1 and S2

C is the carrier to noise in B_{if}.

M is 5

B_{if} is 21 kHz

Ba is 3 kHz

Table 4-3

Print-Out of the Computer Program of Table 4-2
for the Example of Section 4.11.1.

(Headers are defined in Table 4-2.)

S1	S2	D	C
20.42	5.4027	31.749	1
21.42	7.679	23.665	2
22.42	10.109	17.027	3
23.42	12.76	11.642	4
24.42	15.707	7.4363	5
25.42	18.983	4.4028	6
26.42	22.456	2.4914	7
27.42	25.636	1.5079	8
28.42	27.91	1.1245	9
29.42	29.334	1.02	10
30.42	30.412	1.0019	11
31.42	31.42	1.0001	12
32.42	32.42	1	13
33.42	33.42	1	14
34.42	34.42	1	15
35.42	35.42	1	16
36.42	36.42	1	17
37.42	37.42	1	18
38.42	38.42	1	19
39.42	39.42	1	20
40.42	40.42	1	21
41.42	41.42	1	22
42.42	42.42	1	23
43.42	43.42	1	24
44.42	44.42	1	25
45.42	45.42	1	26
46.42	46.42	1	27
47.42	47.42	1	28
48.42	48.42	1	29
49.42	49.42	1	30

Table 4-4.

Computer Program of the Calculation and Plot of *S/N* as a Function of
the Carrier to Noise Ratio and Phase Modulation Index

(The plot is shown in Fig. 4-16.)

```
10 GCLEAR
20 SCALE 0,40,0,60
30 XAXIS 0,10
40 YAXIS 0,10
50 FOR I=0 TO 60 STEP 10
60 XAXIS I
70 NEXT I
80 FOR K=0 TO 40 STEP 10
90 YAXIS K
100 NEXT K
110 LDIR 0
120 FOR X=10 TO 30 STEP 10
130 MOVE X+.5,2
140 LABEL VAL$(X)
150 NEXT X
160 LDIR 0
170 FOR Y=10 TO 50 STEP 10
180 MOVE 2,Y
190 LABEL VAL$(Y)
200 NEXT Y
210 SHORT T1,T2,D,C
220 MOVE 0,0
230 DISP "ENTER BIF,BA,M"
240 INPUT B1,B2,M
250 FOR C1=0 TO 30 STEP 1
260 C=10^(C1/10)
270 S1=M^2*(B1/(B2*2))*C
280 D=1+.9*(B1/B2)^2*(C*EXP(-C))
    /(1-EXP(-C))^2
290 S2=S1/D
300 T1=10*LGT(S1)
310 T2=10*LGT(S2)
320 DRAW C1,T2
330 NEXT C1
340 END
```

4.12 Energy per Bit to Noise Spectral Density (E_b/N_o)

In digital or quasi-analog situations it is often necessary to determine the signal to noise ratio (S/N) relationship to the bit error rate (BER).

The standard method used establishes S/N first by measuring the signal e_s and the noise e_n using a true RMS voltmeter. The true RMS voltmeter must have a sufficient bandwidth to include all of the signal and noise components within the system bandwidth. Having established e_s/e_n in dB it is necessary to add a value k(dB) for conversion to E_b/N_o.

E_b is the energy per bit which is equal to the power level of that bit.

or

$$E_b = \frac{e_s^{\,2}}{z} \cdot \frac{1}{bit\ rate,\ (b_r)} \tag{4-35}$$

where

z is the impedance across which e_s is measured.

Similarly

$$N_o = \frac{e_n^{\,2}}{z} \cdot \frac{1}{ENB} \tag{4-36}$$

where

ENB is the effective noise bandwidth.

The ratio of E_b/N_o is

$$\frac{E_b}{N_o} = \frac{e_s^{\,2}}{e_n^{\,2}} \cdot \frac{ENB}{b_r} \tag{4-37}$$

In $\log_{10}$ notation:

$$\frac{E_b}{N_o}\ (dB) = 20 \log \frac{e_s}{e_n} + 10 \log \frac{ENB}{b_r} \tag{4-38}$$

$$\frac{E_b}{N_o}\ (dB) = k + 20 \log \frac{e_s}{e_n} \tag{4-39}$$

where

$$20 \log \frac{e_s}{e_n} = \frac{S}{N}\ ,\ (dB)$$

and

$$k = 10 \log \frac{ENB}{b_r}$$

or

$$k = ENB - b_r,\ in\ dB.$$

or

$$k = ENB - b_r, \text{ in } dB.$$

Example 4-5:

Let

b_r = 2400 bits per second (bps)
ENB = 2730 Hz
k = 34.36 – 33.8 = 0.56 dB

This value k must be added to each of e_s/e_n measured (dB).

4.13   ERROR FUNCTION (erf)

By definition:

$$\text{erf}(x) = \frac{2}{\sqrt{\pi}} \int_0^x e^{-t^2} dt \tag{4-40}$$

The limiting condition $x \to \infty$ results in erf(x) = 1. Thus we may say $0 \le \text{erf}(x) \le 1$ as $0 < x < \infty$.

Tables of erf(x) are available for use (see reference [3]).

Solutions of erf(x) may be computed for small x (< 2) by the use of the Maclaurin series.

$$\text{erf}(x) = \frac{2}{\sqrt{\pi}} \left(x - \frac{x^3}{1!3} + \frac{x^5}{2!5} - \frac{x^7}{3!7} + \cdots \cdots \right) \tag{4-41}$$

Larger values of x require the use of the asymptotic expansion of $erf(x)$ which is:

$$\text{erf}(x) \approx 1 - \frac{1}{\sqrt{\pi}} e^{-x^2} \left(\frac{1}{x} - \frac{1}{2x^3} + \frac{1 \cdot 3}{2^2 x^5} - \frac{1 \cdot 3 \cdot 5}{2^3 x^7} + \cdots \right) \tag{4-42}$$

For large x

$$\text{erf}(x) \approx 1 - \frac{1}{\sqrt{\pi} x} e^{-x^2} \tag{4-43}$$

A second variation of $erf(x)$ for large x is:

$$\text{erf}(x) \approx 1 - 2k(\sqrt{2} x) \tag{4-44}$$

Where $k(a)$ is

$$k(a) \approx \frac{1}{\sqrt{2\pi} a} e^{-a^2/2} \tag{4-45}$$

and

$$a = \sqrt{2} x \tag{4-46}$$

Example 4-6:

let

x = 2

using equation (4-44)

then

$$\text{erf}(x) = 1 - 2k(\sqrt{2} \cdot 2)$$
$$= 1 - 2k(2.828)$$

and

$$k(2.888) \approx \frac{1}{\sqrt{2\pi \cdot 2.828}} e^{-2.828^2/2}$$

$$\approx .141 \cdot .01833 \approx .00259$$

$$\therefore \text{erf}(x) \approx 1-2(.00259) = 0.99482$$

This compares with

Example 4-7:

Let

x = 2, using equation (4-43)

$$\text{erf}(2) = 1 - \frac{1}{\sqrt{\pi \cdot 2}} e^{-4}$$

$$= 0.994833$$

erf(2) = 0.995322 given by table reference.

It should be noted that the use of the asymptotic series solution does not necessarily result in decreasing error as the series is lengthened, since the solution is cyclic. Where high accuracy is required, the tables are suggested.

4.14 COMPLIMENTARY ERROR FUNCTION (erfc)

The complimentary error function erfc(x) is given by:

$$\text{erfc}(x) = 1 - \text{erf}(x) \qquad (4-47)$$

The solution of erf(c) is explained in detail in 4.13. The compliment of this value is found first by finding erf(x).

4.15 TANGENTIAL SIGNAL SENSITIVITY (TSS)

The TSS definition resulted from the visual observation of a pulse in the presence of noise. When the top of the pulse contains noise, and when the pulse is of such magnitude that the bottom of this noise coincides with the top of the no signal noise, the TSS point has been reached. Fig. (4-17) serves to illustrate this condition. Since a visual observation requires an operator's judgement, some variability in measurement results.

It is generally accepted that TSS corresponds to an output signal to noise ratio of 8 dB which is a power ratio of 6.3 or a voltage ratio of 2.5. Referring this to the input, the detector characteristics must be considered. With seldom-used linear detection, the output signal to noise ration can be transferred directly to the input. This is not the case with square law detectors. To produce an output signal to noise ratio of 2.5 (voltage), its input ratio is $2.5^{1/2}$ which when referred to the input in dB is 4 dB.

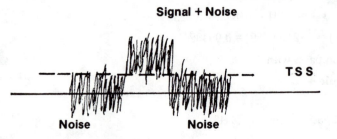

Signal + Noise

T S S

Noise   **Noise**

Fig. 4-17. An illustration of the definition of T S S and the display obtained on an oscillograph.

4.16 CASCADE NOISE FIGURE

Two or more stages, each with its own internally generated noise, when connected in series, will contribute to the overall noise of that group. The overall noise factor of n stages connected in series is described by:

$$F = F_1 + \frac{(F_2 - 1)}{G_1} + \frac{(F_3 - 1)}{G_1 G_2} + \cdots \cdots \frac{(F_n - 1)}{G_1 G_2 G_3 \cdots G_n} \qquad (4\text{-}48)$$

where

    F is the overall noise factor
    F_1 is the noise factor of the first stage (input)
    F_2 is the noise factor of the second stage
    F_3 is the noise factor of the third stage
    F_n is the noise factor of the nth stage
    G_1 is the power gain of the first stage
    G_2 is the power gain of the second stage
    G_3 is the power gain of the third stage
    G_n is the power gain of the nth stage

Note: all values are ratios and are not in dB.

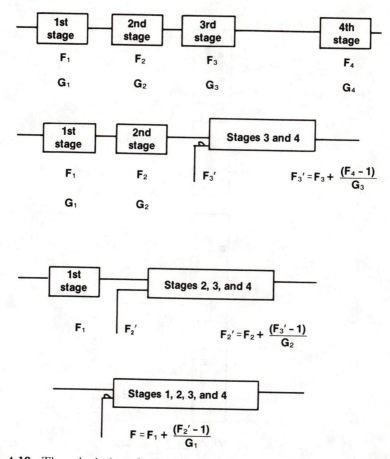

Fig. 4-18. The calculation of noise factor by pairing, which allows the examination of individual stage effects.

When computing the overall noise figure which is $10 \log (NF)$, this calculation becomes cumbersome and does not allow a detailed examination of each stage. In cases where a particularly low noise figure is desired, it is helpful to know the effect of each stage.

A more practical calculation results by pairing and using successive application of the noise factor calculation for two stages, starting with the last stage. This process is shown in Fig. (4-18).

An example showing the calculation of the overall noise figure of a receiver, using the pairing of stages, is shown in Fig. (4-19). Note that attenuation adds directly to the noise figure of the following stage. That is, the noise figure of the first IF amplifier is increased by the noise figure of the IF crystal filter and that

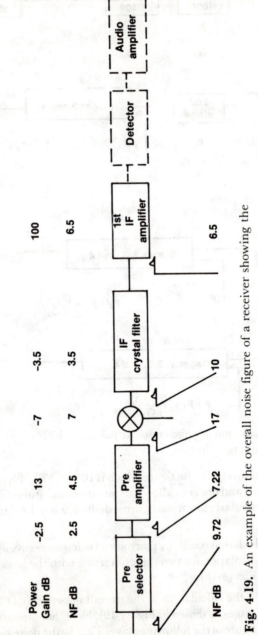

Fig. 4-19. An example of the overall noise figure of a receiver showing the effects of each stage.

of the passive mixer in a simple sum. This calculation is simply $6.5 + 3.5 + 7 = 17$ dB. This fact reduces the computation task, which may be done by inspection up to that point.

The computer program of Table 4-5 is useful for the calculation and analysis of cascaded stages. The user need only answer the input requirements. The program begins with the last stage and proceeds toward the input stage.

Table 4-5.

Computer Program for the Calculation of Cascade Noise Figure

```
10 PRINT " CASCADE NOISE FIGURE
   "
20 PRINT " STARTS WITH LAST STA
   GE AND WORKS UP TO THE INPUT
   "
30 PRINT "  NFT=10*LGT(F1+((F2-
   1)/G1));DB"
40 PRINT "  WHERE"
50 PRINT " NFT=TOTAL NOISE FIGU
   RE(DB)"
60 PRINT " F1=PRECEDING STAGE N
   OISE FIGURE(RATIO)"
70 PRINT " F2=FOLLOWING STAGE N
   OISE FIGURE(RATIO)"
75 PRINT " G1=1ST STAGE GAIN,DB
   "
80 PRINT "ALL PROGRAM ENTRIES A
   RE IN DB"
90 PRINT
100 PRINT "*************************
    ***********"
110 PRINT "NF     G     CAS NF
         STAGE     "
120 PRINT "DB        DB        DB "
130 PRINT
140 DISP "IDENT LAST STAGE(<11 C
    HAR)"
150 DIM A$[15]
160 INPUT A$
170 DISP "ENTER LAST NF,GAIN(DB)
    ";
180 INPUT F2,G2
190 PRINT TAB(1);F2;TAB(7);G2;TA
    B(22);A$
210 DISP "IDENT NEXT STAGE(<11 C
    HAR)";
```

```
220 DIM B$[15]
230 INPUT B$
240 DISP "IDENT NEXT STAGE NF,GA
    IN(DB)";
250 INPUT F1,G1
260 X=10^(F1/10)
270 U=10^(F2/10)
280 V=10^(G1/10)
290 W=10*LGT(X+(U-1)/V)
295 Y=INT(W*10+.5)/10
300 PRINT TAB(1);F1;TAB(7);G1;TA
    B(13);Y;TAB(22);B$
310 F2=W
320 GOTO 210
330 END
```

4.17   INTERMODULATION DISTORTION (IM)[4]

When two signals whose frequencies F_1 and F_2 are applied to a non-linear device, other frequencies are generated at:

$$F_1' = F_1 - F\Delta$$

and

$$F_2' = F_2 + F\Delta$$

where

$$F\Delta = F_2 - F_1 \text{ and } F_2 > F_1$$

This shown in Figure (4-20).

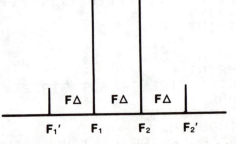

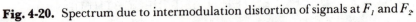

Fig. 4-20. Spectrum due to intermodulation distortion of signals at F_1 and F_2.

Assume that a receiver is tuned to a frequency $F_3 = F_2'$. Whenever signals appear simultaneously at frequencies F_1 and F_2, an unwanted signal F_2' appears at F_3, which for all practical purposes appears to be a legitimate signal. This spurious signal may cause serious interference for a desired signal.

Consider a 1000 channel system with 25 kHz channel spacing. Intermodulation products can result for any pair of signals whose separation from each other is $N \cdot F\Delta$ and they are spaced $N \cdot F\Delta$ and $2 NF\Delta$ respectively, from a desired channel, where $F\Delta = 25$ kHz. The possibilities of interference resulting from intermodulation distortion are very great.

All signal processors are non-linear to some degree and will cause the generation of intermodulation distortion terms.

When two signals

$$[A_1 \cos w_1 t \text{ and } A_2 \cos (w_2 t + \delta)] = \rho_s \tag{4-49}$$

are simultaneously applied to a non-linear processor whose transfer function is of the form

$$K_1 \rho_s + K_2 \rho_s^2 + K_3 \rho_s^3 + \cdots Kn \rho_s^n \tag{4-50}$$

output terms result which include all possible products. The most significant of these are

$$K_1 [A_1 \cos w_1 t + A_2 \cos (w_2 t + \delta)] \tag{4-51}$$

which represents the two signals,

$$0.5 A_1^2 K_2 \cos 2w_1 t$$

$$0.5 A_2^2 K_2 \cos 2 (w_2 t + \delta)$$

$$A_1 A_2 K_2 \cos [w_1 t - w_2 (t + \delta)]$$

$$A_1 A_2 K_2 \cos [w_1 t + w_2 (t + \delta)]$$

representing the second order terms and

$$0.75 K_3 A_1^2 A_2 \cos [2w_1 t \pm w_2 (t + \delta)]$$

$$0.75 K_3 A_1 A_2^2 \cos [2w_2 (t + \delta) \pm w_1 t]$$

which represent the third order terms of interest.

Since these products have different slopes, they will cross at some point for a given input level. This intersection is termed the intercept point (Fig. (4-21)). Once this point is determined it is possible to predict the intermodulation product magnitudes, with good accuracy. The charts of Fig. (4-22) and (4-23) may be used to obtain the absolute or relative magnitudes of the second or third order products, given the second or third order intercept point, respectively.

It is also possible to calculate the distortion product magnitudes by noting that for a 0 dBm two-tone signal input, the third order intercept is equal to $-1/2$ of the third order product magnitude for a unity gain device. For example, given a third order intercept point of +20 dBm, the third order distortion product magnitudes would be $20 \div (-1/2) = -40$ dBm, for a two-tone input signal with 0 dBm magnitude. This relationship may also be used in reverse. Given a third order product magnitude of –60 dBm for 0 dBm two-tone input signal and unity gain device, the intercept point is –60 × (–1/2) = 30 dBm.

All intercept information is referenced to the output unless otherwise specified. This includes the two-tone magnitude as well as the distortion product level. Assume an amplifier has gain of 10 dB and the two-tone input magnitude is –10 dBm. The amplifier output will consist of the two-tone signals whose magnitude is now 0 dBm; assuming that the third order products are –40 dBm, the third order output intercept point is +20 dBm. To relate this output intercept value to the input, simply subtract the gain.

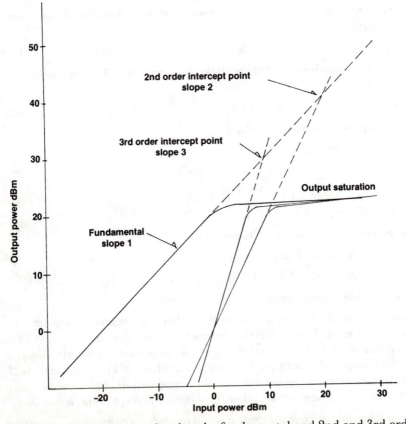

Fig. 4-21. Device outputs showing the fundamental and 2nd and 3rd order distortion products together with the extrapolated respective intercept points.

When using other than 0 dBm two-tone signals, normalize the levels to 0 dBm remembering that the third order distortion products increase 3 dB for every dB increase in the two-tone level.

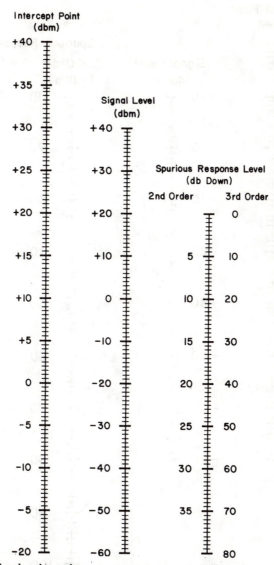

Fig. 4-22. Relative level spurious response nomograph. (Based on nomographs from Avantek, Inc., Santa Clara, California.)

For a two-tone level of –30 dBm, at the input of an amplifier whose gain is 20 dB, we have an output two-tone level of –10 dBm. Assume that the third order distortion products have a magnitude of –50 dBm. To normalize the two-tone output signal level of –10 dBm, add 10 dBm. Also add 3 · 10 dBm to the third order products (–50 + 3 · 10 = –20 dBm). The output third order intercept point

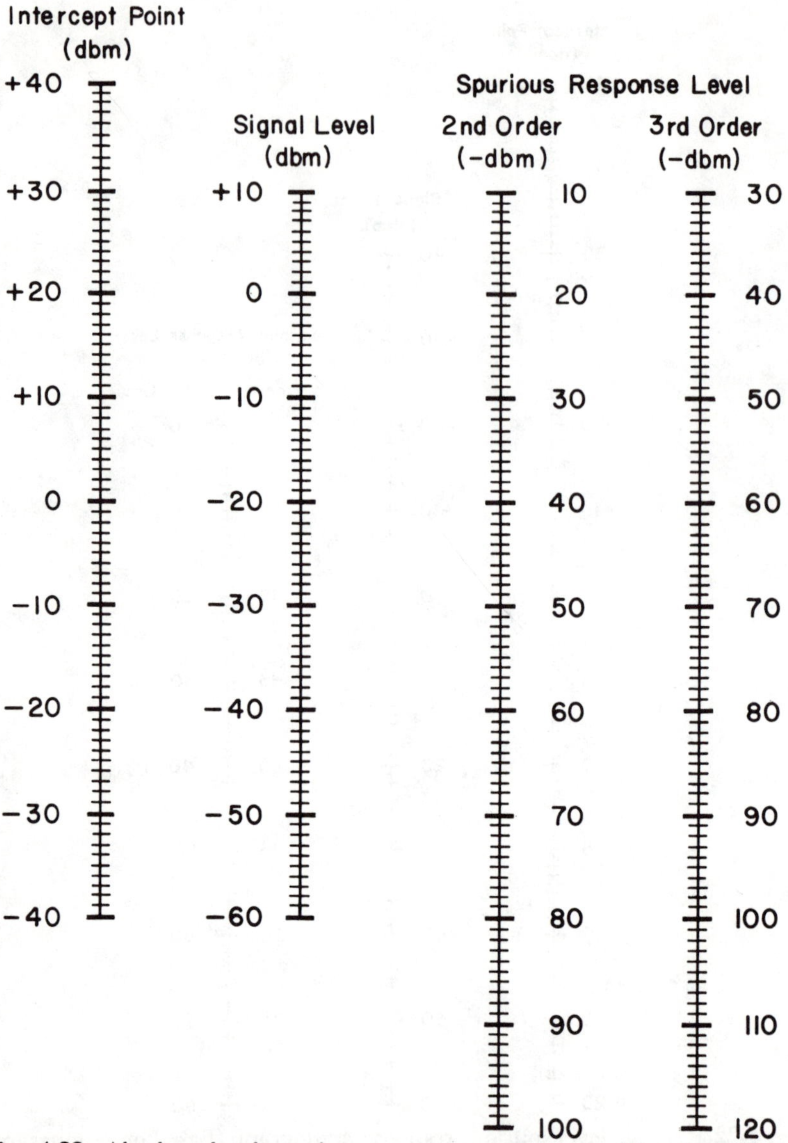

Intercept Point
(dbm)

+40

+30   Signal Level   Spurious Response Level
                         (dbm)   2nd Order   3rd Order
+30   +10   (−dbm)   (−dbm)
                                              10   30

+20   0   20   40

+10   −10   30   50

0   −20   40   60

−10   −30   50   70

−20   −40   60   80

−30   −50   70   90

−40   −60   80   100

                                              90   110

                                             100   120

Fig. 4-23. Absolute level spurious response nomograph. (Based on nomographs from Avantek, Inc., Santa Clara, California.)

is −20 dBm · (−1/2) = 10 dBm.

When dealing with relative magnitudes, proceed as above, except note that there is a 2 dB/dBm relationship between the output two-tone signals and the distortion products.

For the case where the two-tone signals are unequal in magnitude, simply subtract $1/3$ the difference between them from the larger.

Given:

Signal 1   +18 dBm
Signal 2   0 dBm

The equivalent equal magnitude two-tone signal has a power level of $+18$ dBm $-1/3$ (18 dBm -0 dBm) = 12 dBm. Then proceed as before, using either the charts of calculation.

Second order intercepts are seldom considered because those products are generally farther removed from the desired frequency. Should it be of interest, second order terms may be related by the intercept point, as before.

4.17.1 Cascade Intercept Point

The system designer may be called upon to predict the intermodulation performance of many stages in cascade. One such application may involve a receiver where the intermodulation performance is specified. Assume that the specification reads: "The intermodulation distortion products resulting form a two-tone -30 dBm signal, seperated 2 and 4 MHz (respectively) from the desired frequency, and on the same side, shall not exceed -100 dBm." From Fig. (4-22) or (4-23), or by calculation, the third order intercept point must not be less than $+5$ dBm at the receiver input.

Further, assume that the receiver of Fig. (4-24) is being considered for this application. Since the two tones require 4 MHz of bandwidth, they will be processed by all stages, up to the first IF filter, which has a 50 kHz bandwidth and does not allow them to pass. This is the intermodulation distortion truncation point. All calculations must include all stages between the antenna and this filter which heavily attenuates the two-tone signal. Further intermodulation distortion contributions are negligible.

Norton's equation (4-52) may now be applied successively stage by stage, beginning at the antenna, up to the determined truncation point and the output intercept point determined. The gain to this point must be subtracted from this value to obtain the input intercept point, which may now be compared to the requirements.

$$I_t^3 = I_2^3 - 10 \log \left[1 + \frac{1}{g_2} \cdot \frac{I_2^3}{I_1^3} \right] \qquad (4\text{-}52)$$

where

$\overset{3}{I_t}$ is the third order cascade output intercept point (dBm),

$\overset{3}{I_2}$ is the second stage third order output intercept point (dBm),

g_2 is the power gain of the second stage, and

$\overset{3}{I_t}$ is the first stage third order output intercept point.

Note: The terms in the brackets are not in dB notation. Use the numerical equivalent. (See reference [5]).

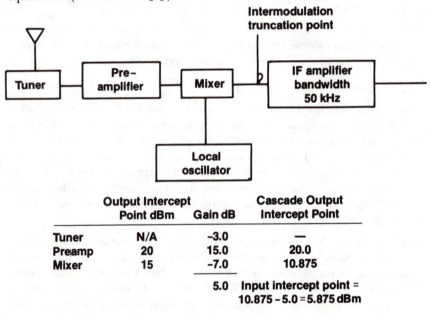

	Output Intercept Point dBm	Gain dB	Cascade Output Intercept Point
Tuner	N/A	–3.0	—
Preamp	20	15.0	20.0
Mixer	15	–7.0	10.875
		5.0	Input intercept point = 10.875 – 5.0 = 5.875 dBm

Fig. 4-24. Receiver example showing the input intercept point calculation.

The previous simplified example illustrates the cascade calculation method which indicates an input intercept point of 10.875 – 5 = 5.875 dBm. This numerical value barely meets the requirement. However, it must be kept in mind that Norton's equation assumes coherence of the intermodulation products through the various stages, which is not likely. The calculation is therefore considered pessimistic.

The second order intercept point may be computed in similar fashion using the equation:

$$\overset{2}{I_{t}} = \overset{2}{I_{2}} - 20 \log \left[1 + \sqrt{\frac{1}{g_{2}} \cdot \frac{\overset{2}{I_{2}}}{\overset{2}{I_{1}}}} \right] \tag{4-53}$$

$\overset{2}{I_{t}}$ is the second order cascade output intercept point (dBm)

$\overset{2}{I_{2}}$ is the second stage second order output intercept point (dBm)

g_{2} is the power gain of the second stage

$\overset{2}{I_{1}}$ is the first stage second order output intercept point

Note: the terms in the brackets are not in dB notation.

The procedure is to begin at the input as the first stage and compute the cascade with the following stage. This value becomes the first stage and the next stage (third in this case) becomes the next or second stage, *et cetera*.

The input intercept point becomes the final cascade intercept point minus the preceding gain in dB notation, as shown in Fig. (4-24). This computation can be tedious for complex systems and it is suggested that the program of Table 4-6 be utilized for that purpose.

Table 4-6.
Computer Program for the Calculation of Cascade Intercept Point

```
 10 PRINT "   CASCADE INTERCEPT"
 20 PRINT "   COMPUTES DEGRADATIO
    N OF THE INTERCEPT POINT DUE
     TO A PRECEDING STAGE"
 30 PRINT
 40 PRINT "   THE INPUT INT PT IS
     THE OUTPUT INT PT-GAIN"
 50 DISP "CHOOSE 2ND (2) OR 3RD
    (3) ORDER";
 60 INPUT N
 70 PRINT "CASCADE INTERCEPT POI
    NT"
 80 IF N=2 THEN 100
 90 IF N=3 THEN 120
100 PRINT "SECOND ORDER"
110 GOTO 130
120 PRINT "THIRD ORDER"
130 PRINT "***********************
    ***********"
140 PRINT "IPT  G  CAS I  GT  IN
    IPT  STAGE"
150 PRINT "DBM  DB  DBM    DB   D
    BM"
```

```
160 PRINT
170 DISP "TITLE 1ST STAGE(5CHAR)
    "
180 DIM A$[5]
190 INPUT A$
200 DISP "ENTER 1ST STAGE INT PT
    (DBM)"
210 INPUT X
220 DISP "ENTER 1ST STAGE GAIN(D
    B)";
230 INPUT P
240 PRINT TAB(1);X;TAB(6);P;TAB(
    28);A$
250 PRINT
260 DISP "TITLE 2ND STAGE(5 CHAR
    )";
270 DIM B$[5]
280 INPUT B$
290 DISP "ENTER 2ND INT PT(DBM),
    GAIN(DB)";
300 INPUT Y,Z
310 K=20/(N-1)
320 L=10^(Z/10)
330 M=10^(Y/10)
340 J=10^(X/10)
350 D=P+Z
360 C=INT((Y-K*LGT(1+(1/L*M/J)^(
    1/(4-N))))*10+.5)/10
370 E=C-D
380 PRINT TAB(1);Y;TAB(6);Z;TAB(
    10);C;TAB(15);D;TAB(20);E;TA
    B(28);B$
390 PRINT
400 X=C
410 P=D
420 GOTO 260
430 END
```

The most frequent error in the computation of the second order intercept point is the determination of the truncation point. To determine this point, it is first necessary to develop an understanding of the second order distortion terms. These have been shown to be of the form:

Second harmonic terms, $2F_1$, $2F_2$

Sum and difference terms:

$F_1 \pm F_2$

Where

F_1 and F_2 are the two frequencies which cause the second order distortion.

There can be two cases which must be considered.

These are:

In band and
Out of band

The in band case results when F_1 and F_2 are within the receiver's tuning range. The out of band case results when these two signals are out of the receiver's tuning range.

The only barriers to second order distortion generation are the preselector attenuation, the conversion of the first mixer, and the use of high second order intercept components.

Many of the distortion products will not be converted by the first mixer and the remainder will suffer preselector attenuation. Where the conversion is the truncator the designer need not look beyond this point for further additive distortion. The attenuation of the preselector of the signal F_1 and (or) F_2 serves to increase the systems intercept point, which permits the use of lower intercept point components, if desired. Seldom is it ever necessary to extend the intercept analysis to the detector.

The following is a good approach to second order analysis:

Determine the range of the signals involved
Analyze the effect of the preselector on these signals

Examine the conversion of these distortion terms coordinated with preselector tuning.

A good design will attenuate the signals F_1 and (or) F_2 to livable levels and also reject through conversion the remainder.

Note: A simple relationship permits the determination of intercept point knowing the level of the distortion products.

This relationship is as follows:

$$\overset{m}{I} = S + \frac{\overset{m}{R}}{(m-1)}$$

where

m is the order

S is the signal (dBm)

$\overset{m}{R}$ is the intermodulation ratio (dB)

Example 4-7:

The requirements call for a suppression of second order products of 60 dB, resulting from signals of –50 dBm.

Then

$$\overset{m}{I} = -50 + 60/(2-1) = 10 \text{ dBm}$$

The equation is equally applicable to the third order case.

4.18 DESENSITIZATION

Desensitization of a receiver is the reduction of gain of that receiver to a desired frequency by a second unwanted signal. This can result from the presence of a stronger signal within the IF bandwidth of an FM receiver, the compression of stages in an AM receiver or from AGC takeover in AM and FM receivers, if used for the latter case.

Poorly designed receivers are subject to AGC takeover if the final selectivity is placed before the AGC detector, instead of at the front of the IF amplifier. An example of this is a multibandwidth multifunction receiver which must provide simultaneous outputs at several bandwidths. The filter following the final IF converter must be wider than the widest final filter bandwidth preceding the detectors. The IF strip is exposed to signals in bandwidth window possibly wider than that of the AGC system. The result of this is potential blocking or overload of the receiver.

A second possibility is AGC derived in a wide bandwidth with narrow band detection. Here a strong unwanted signal developes AGC, resulting in a weak desired output. To reduce desensitization effects, the designer must provide proper filtering throughout the receiver and eliminate sources of broadband noise (external and internal) to the receiver.

4.19 COMPRESSION

All linear systems, when given a sufficiently strong input signal, depart from a linear relationship between input and output. Such a system is said to be going into compression or saturation. An example of this situation would be an amplifier whose performance is shown in Fig. (4-25). The performance is linear up to an input signal level in excess of 0 dBm. Beyond this point, the output falls below the linearly extrapolated input/output characteristic and the amplifier is no longer linear.

Compression, or the –1 dB compression point, is generally taken as that point of –1 dB departure from linearity. In the example this is 10 dBm input or 19 dBm output. The input signal level which causes this departure is the value of compression usually specified. Where the output compression point is given, the gain between input and output (in dB) is subtracted from the output, to secure the input compression point.

Non-linear processes which provide a linear output amplitude, such as a mixer, are subject to compression as well. In this case the mixer is operating with a fixed local oscillator drive level and the input signal is increased while monitoring the output. Care must be exercised in this measurement to exclude all but the desired converted signal component. This may be readily accomplished by using a spectrum analyzer and examining the desired spectral line, or by using a RF voltmeter preceded by an appropriate IF filter. As a rule of thumb for double balanced diode mixers, the output compression point is the conversion loss below the local oscillator drive level. The input compression point is the output compression point increased by the amount of the conversion loss through the mixer.

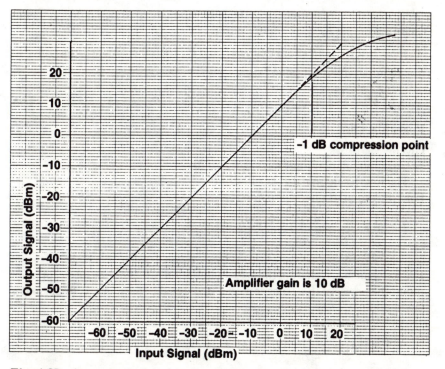

Fig. 4-25. An example showing the compression of a linear amplifier. Where the departure from linearity is –1 dB, this is the –1 dB compression point.

Example 4-8:

Given an output compression point of 9.5 dBm and a conversion loss of 6.5 dB, the input compression point is

9.5 + 6.5 = 16 dBm

4.20   CROSS MODULATION

When two signals appear simultaneously at a receiver's input, and one is modulated and the second is not, non-linearities in the receiver will impart modulation to the unmodulated signal from the modulated one. This process is called cross modulation and it is related to the third order intercept point by

$$
M_c = \frac{\overset{3}{I}}{4P_2} + \frac{1}{2} \tag{4-54}
$$

where

$\overset{3}{I}$ is the receiver third order intercept point

P_2 is the power level of the stronger modulated signal

Note that the signal strength of the lesser unmodulated signal does not enter into the computation.

In most cases specifications define M_c and P_2; then the designer must solve for the required intercept point $\overset{3}{I}$ from

$$
\overset{3}{I} = \left(M_c - \frac{1}{2} \right) 4P_2 \tag{4-55}
$$

Example 4-9:

Given:

 Cross modulation must not exceed 20 dB

 The interfering signal P_2 is –10 dBm

Find the required intercept point.

$$
\overset{3}{I} = \left(\text{anl} \ \frac{20}{20} - \frac{1}{2} \right) 4 \, \text{anl} \ \frac{-10}{10}
$$

   = 5.79 dB

For convenience, a chart of the cross modulation ratio and the intercept point minus the power of the stronger modulated signal is given in Fig. (4-26) and is based upon:

$$
\frac{\overset{3}{I}}{P_2} = 4 \left(M_c - \frac{1}{2} \right) \quad \text{Where } M_c = \frac{m}{m_I} \tag{4-56}
$$

 m is the modulation index

 m_I is the effective modulation index

Example 4-11:

Find the solution to the previous problem using the chart of Fig. (4-26). Enter the chart at $m/m_I = 20$ dB and read $\overset{3}{I} - P_2 = 16$. Since $P_2 = -10$, $\overset{3}{I} = 6$ dBm. References [6] and [4] contain additional information on this subject.

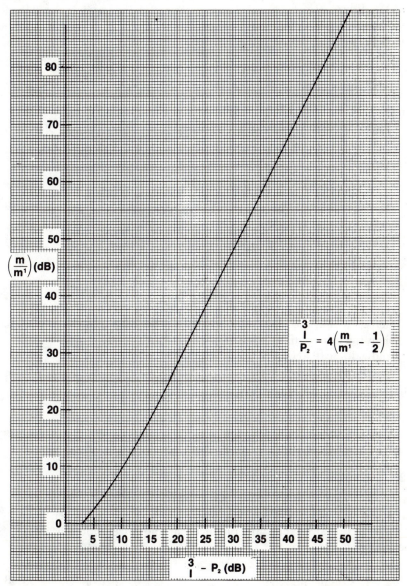

$$\frac{\overset{3}{I}}{P_2} = 4\left(\frac{m}{m^I} - \frac{1}{2}\right)$$

Fig. 4-26. Graph of cross modulation *versus* intercept point [6].

4.21 SPURIOUS FREE DYNAMIC RANGE

It is informative to determine the dynamic signal operation range of a system which is free of spurious signals resulting from third order intermodulation products. In this definition spurious free means that these spurious signals are equal to the noise level. The third order intermodulation product signal levels are related to the level of the two signals (producing them, respectively, on a three for one dB relationship). The spurious free dynamic range is related to the third order system intercept point by the following relationship:

$$SFDR \text{ (dB)} = 0.67 \,(\overset{3}{I} - kT/\text{MHz} - 10 \log B - NF) \qquad (4\text{-}57)$$

where

SFDR is the spurious free dynamic range

$\overset{3}{I_i}$ is the system third order input intercept point
kT is the thermal noise level in a 1 MHz bandwidth
   = –114 dBm / MHz
B is bandwidth in MHz
NF is the system noise figure (dB)

An example of the application of this relationship follows.

Example 4-11:

Given:

Third order input intercept point = 10 dBm (Note: Given the output intercept point, the input intercept point is the output intercept point minus the gain in dB notation.)
Bandwidth is 10 kHz
Noise figure is 5 dB

Find the spurious free dynamic range.

Solution:

SFDR = 0.67 (10 – (–114) – (–20) – 5)
    = 92.67 dB

4.22 IMAGES

In the mixing process it was shown that when two input signals comprised of an RF signal and a local oscillator signal are applied to a mixer, intermediate frequency signals are produced at the mixer's output. More specifically:

$$F_{if} = \left| NF_r \pm MF_{lo} \right| \qquad (4\text{-}58)$$

where

F_{if} is the intermediate frequency
F_r is the receive frequency
F_{lo} is the local oscillator frequency
M and N are intergers

The primary or desired outputs result when $M = N = 1$

Then equation (4-58) results in:

$$F_{if} = \left| F_r \pm F_{lo} \right| \tag{4-59}$$

Since F_{if} is a constant, then for a given LO frequency there exist two values of F_r which satisfy the relationship.

EXAMPLE 4-13:

Let F_{lo} = 160.7 MHz
   F_{if} = 10.7 MHz
then $F_r = \left| F_{if} \pm F_{lo} \right|$
or $\left| 10.7 \pm 160.7 \right|$ = 150 and 171.4 MHz.

Thus, the mixer is equally responsive to two frequencies, both of which are twice the IF apart. Of these, one is the desired response and the other is called the image frequency. The receiver must reject the image term to provide satisfactory performance. This is one of the reasons mixers are always preceded by preselection filtering of some form. The selectivity requirements of the preselector are governed by the frequency separation between the image frequency and the desired frequency. The amount of image frequency rejection is strictly a function of the attenuation, provided by the preselector filter alone (unless an image rejection mixer is used).

To avoid extreme selectivity requirements from the preselector, the ratio of F_r to F_{if} should not exceed 10 or 20 to 1 for a first conversion in a down conversion superheterodyne receiver.

In up-conversion systems, the high value of intermediate frequency effectively removes the image frequency out of the preselector bandwidth. Here, simple fixed tuned filters often suffice as preselector filters, eliminating tracking and tuning problems. Fig. (4-27) is an example illustrating the relative image frequency behavior between up and down conversion.

4.23 HIGH ORDER IMAGES

When a receiver design utilizes more than one conversion, the image problem becomes more complex. For every mixer in a receiver there will exist an image frequency. A double conversion receiver will have primary and secondary

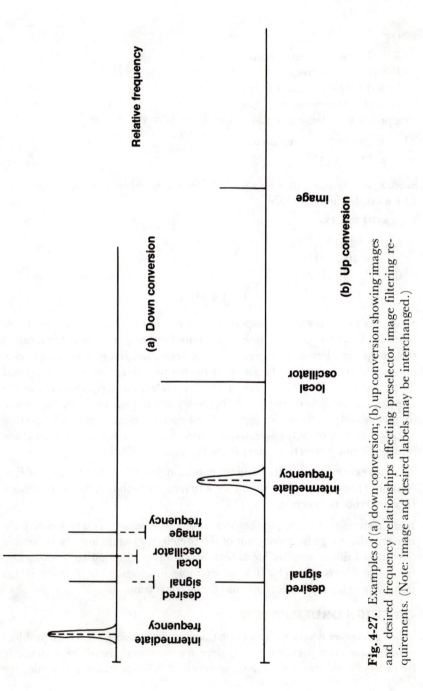

Fig. 4-27. Examples of (a) down conversion; (b) up conversion showing images and desired frequency relationships affecting preselector image filtering requirements. (Note: image and desired labels may be interchanged.)

images. A triple conversion receiver will have primary, secondary, and tertiary images, and so on.

The treatment for high order images is the same as for the primary case. The only means of image rejection is that of the filter, which precedes the mixer. For this reason, in multiple conversion systems, the IF filter preceding the mixer in second and higher conversions must provide attenuation at the respective image frequency. In some cases, additive attenuation may be provided by filters and/or preselection from earlier conversions. For example, a secondary image may also be attenuated by the preselector. A tertiary image could be rejected by the second IF filter, *et cetera*. In difficult cases other sources of attenuation may be of value.

To illustrate the situation, an example of a triple conversion receiver is provided in Fig. (4-28). To illustrate the situation, the parameters used in the illustration are not necessarily optimized for a real-world design.

The higher the order of an image frequency the more stages are involved, and the likelihood of securing additional attenuation is generally greater. In the illustration, had the third IF been lower and the receiver tuning range greater, the tertiary image would have been in band and the preselector would not have had any effect in the worst case.

A summary of the attenuators for the respective images follows.

Primary Image

The attenuation is the ultimate attenuation of the preselector at the image frequency. In this case the image frequency is far removed from the preselector tuning range and the ultimate attenuation may rebound from its floor value. For this reason the preselector must be specified, designed and tested at the image frequency to ensure predictable performance. For this illustration a value of −70 dB is assumed. In some cases this may not be sufficient and an image rejection (lowpass) filter would be added ahead of mixer M_1.

Secondary Image

The preselector provides its ultimate attenuation to the secondary image frequency, (the previous comments made in the primary image's case still apply.) Additionally, the first IF filter will provide attenuation. For a four-section Chebyshev filter with a 0.1 dB ripple the value is 95.25 dB. This is well beyond the usual ultimate attenuation, so a value of 70 dB will be assumed. The total attenuation is 140 dB.

Tertiary Image

The rejection of the tertiary image frequency is the sum of the attenuation of

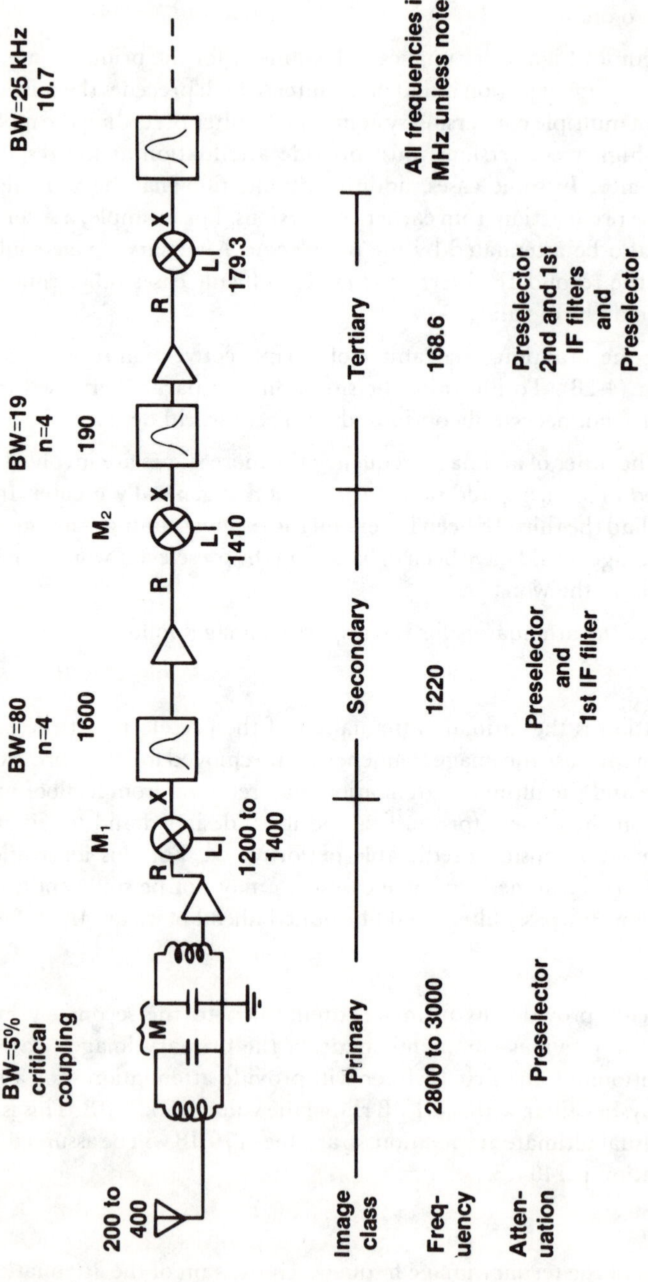

Fig. 4-28. An illustration of secondary and third order images. (Broad band amplifiers and mixers were assumed.)

the preselector, plus the first and second IF filters at the image frequency.

An attenuation value of 33.7 dB is computed for the second IF filter. The first IF filter supplies its ultimate value of 70 dB by inspection (large frequency offset). The preselector is assumed to have the following characteristics:

Unloaded Q
$Q_u = 90$
Loaded Q
$Q_l = 200/10 = 20$

Since the tertiary image frequency is 168.6 MHz, the worst case preselection attenuation results when the preselector is tuned to 200 MHz. A bandwidth of 5% at 200 MHz is 10 MHz.

Therefore

Q_l = center frequency/bandwidth

Using Table 5-4 for critically coupled tuned circuits, the preselector is found to provide 11.4 dB of attenuation.

The total tertiary image rejection due to filtering alone is:

33.7 + 70 + 11.4 = 115.1 dB

In real situations, additional sources of attenuation are provided by mixers and amplifiers. For example, it would not be economical to use a second IF amplifier (which would be so broad as to pass the tertiary image frequency) or for that matter any frequency other than those required. Mixers are also relatively narrowband. They range from narrowband to several octaves at high frequencies, or to more than a decade at low frequencies.

In an image performance analysis the response of every stage should be considered. Often, after having determined the required rejection the analysis may stop at a point, providing the minimal attenuation value plus margin.

As a typical case, a specification may define a signal environment of –10 dBm, representing the maximum value. Signals of this magnitude could be encountered at the image frequencies. The specification may also dictate that image responses must be equal to noise. Assume that in this case it is found that the noise floor is –120 dBm. The required image attenuation must be –120 – (–10) = –110 dB minimum.

For the system illustrated, it can be seen that the primary image rejection is only 70 dB and an additional 40 dB-plus margin must be secured. In a real case, the preamplifier would usually not pass the primary image frequency and additional attenuation would be provided by this stage. In addition, the mixer's R port response may be down, providing more primary image attenuation. All of these sources should not be overlooked in difficult cases. Should the image rejection still be inadequate, then additional filtering could be added at the primary image frequency. Having completed the analysis and made the necessary circuit refinements the image performance predictions give confidence to the design adequacy.

The previous discussion assumes that the image frequencies are applied at the receiver input connector and many specifications may so state. Although meeting the letter of the design specification is legally correct it is not ethical to overlook internal-image frequency generation, even though no specification may cover the case. In Fig. (4-28) it can be seen that the first local oscillator ranges from 1200 to 1400 MHz and the secondary image frequency is 1220 MHz. When the receiver is tuned to 380 MHz, the first local oscillator will generate the secondary image frequency of 1220 MHz. This can be a serious problem.

Modern high performance receivers may utilize first local oscillator power levels of 20 dBm or more. Assuming that this is the case, the secondary image frequency signal level at the X port of the first mixer. M_1 would be 20 –70 (ultimate attenuation of the first IF) – 20 (mixer L to X *port isolation*) = –70 dBm at the first IF filter output. This level must be equal to noise level at the IF filter output to meet specifications. The noise level at the first IF filter output is the input noise level at the attenna connector plus the intervening gain (or loss). In Fig. (4-28) assume:

Preselector loss	2.18 dB
Preamplifier gain	15 dB
Mixer M_1 conversion	– 7 dB
First IF filter	– 3 dB
total	2.82 dB

The input noise level was –120 dBm. Therefore the noise level at the output of the first IF filter is –120 dBm + 2.82 dB = –117.18 dBm. To meet the specifications the magnitude of the image frequency from the local oscillator must be less than –117.18 dBm. The required additional filtering must be:

   –117.18 dB – (–70 dBm) = 47.18 dB min

An additional filter could be added following the first IF amplifier, so that the requirements are met. Here it was assumed that the first IF amplifier was equally responsive to the first IF frequency and the secondary image frequency. Where this is not the case, the first IF amplifier attenuation would contribute to the requirements.

The best solution is to avoid such situations by relocating the frequencies involved. Here the second local oscillator frequency could have been moved to 1790 MHz, thereby shifting the secondary image frequency to 1980 MHz.

This case was used to illustrate the pitfalls in analysis. It is good practice to examine the design for internal sources of image frequencies and take necessary action.

4.24 SELECTIVITY

The selectivity of a superheterodyne receiver is determined solely by the selectivity of the IF amplifier. Here the selectivity may be obtained by the use of selective networks such as tuned circuits, crystal filters, or both.

The need for selectivity results when several signals are presented to the receiver simultaneously. Assuming that these signals are in close proximity to the desired signal, as in adjacent channels in a channelized network, the receiver must reject all but the desired channel.

A typical communications network will assign specific frequencies to each included station, and each frequency will be spaced from the frequencies of adjacent stations by a fixed amount. In a high density environment, and for maximum utilization of the RF spectrum, this spacing is minimized and determined by:

the spectral width of the transmitted signal
the frequency inaccuracies of the transmitter
the frequency inaccuracy of the receiver
plus margin

The situation is illustrated in Fig. (4-29).

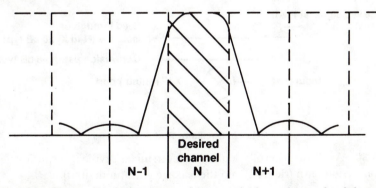

Fig. 4-29. Channelized spectrum with a typical receiver selectivity curve superimposed.

The receiver selectivity curve's solid line is shown to include the desired channel (N) within a reasonably flat part of its response. Being non-ideal, the response falls off but is not totally exclusive to channels $N+1$ and $N-1$, *et cetera*. Therefore, signals within these adjacent channels could cause interference to those of the desired channel N and under certain conditions cause complete communications failure.

Frequency management avoids this problem by allowing guard bands of one or more channels on either side of a channel within a geographical zone. This zone is determined by a signal intensity profile which considers path losses, transmitted power and antenna directivities, to ensure that adjacent channel signal intensity is low enough to minimize serious interference.

The selectivity of the filter is not usually achievable at the generally high receive frequency and is therefore applied at a lower intermediate frequency.

From this discussion, we develop the following rule:

> To ensure optimum selectivity, it is good design practice to restrict the receiver IF bandwidth to that of the received signal's spectral width plus the sum of the drift allowance for both the transmitter and receiver. Additionally, the shape of the filter should be made as nearly rectangular as possible, constrained only by permissible distortion.

4.25 INTERMEDIATE FREQUENCY (IF) REJECTION

A mixer will pass signals present at its input (R port) to its output (IF port), without conversion through its imperfect isolation, between these two ports. This is illustrated in Fig. (4-30). Typically, this isolation value for a double balanced mixer is 20 dB. This unwanted signal path can result in desensitization and heterodyne problems, or both.

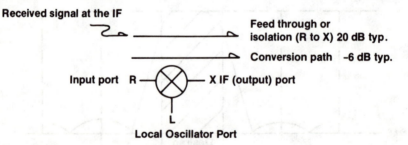

Fig. 4-30. An illustration of a double balanced mixer's typical output through the conversion path (desired) and the leakage path (undesired).

Since it was shown in section 4.5 that the IF must be out of band, the only other source of IF signal attenuation is presented by the preselector filter's and preamplifier's frequency response.

The designer must provide a preselector and preamplifier (if used) response which, when added to the isolation of the mixer, results in a IF rejection level which meets the specifications.

Example 4-14:

Specification:

When the receiver is presented with a signal level of –25 dBm at the IF, the resulting receiver output shall not exceed the noise level.

For a receiver with a noise level of –110 dBm, the receiver front end must provide an attenuation greater than 65 dB.

$$|-110-(-20 -25)| = 65 \text{ dB min where the } L \text{ to } R \text{ isolation is 20 dB.}$$

Achieving IF rejection is generally not difficult unless the designer selects an IF too close to the received frequency.

4.26 LOCAL OSCILLATOR RADIATION

A local oscillator (LO) or any oscillator is a transmitter if provided with a suitable radiator, conduction, or leakage path. This can be troublesome within a receiver itself or to other co-located receivers.

Internal oscillator problems within a receiver are a design problem and can be cured by appropriate shielding, isolation, layout *et cetera*. The co-location problem can be severe within an installation and for this reason specifications may be applied which govern the amount of local oscillator radiation that can be tolerated.

A typical specification may dictate that the local oscillator signal present at the receiver antenna connector be equal to or less than noise by x dB. It may also specify in band and out of band radiation. The design treatment in any case is the same.

From the discussion of mixers it was shown that even with the best double balanced mixers available, local oscillator to receive port isolation is only 20 to 30dB. Assuming a LO power level of 10 dBm, a signal due to the local oscillator will appear at the mixer input receive port. This signal, with a magnitude of 10 – (20 to 30) = –10 to –20 dBm, would cause severe interference to other receivers in the vicinity if it were coupled to an antenna and tuned to this frequency. This This problem is particularly severe with down conversion receivers (low IF), where the local oscillator is the IF away from the tuned frequency.

To reduce local oscillator radiation, preselection and preamplification are employed. The local oscillator radiation requirements generally dictate the selectivity requirements of a preselector, more so than image rejection. This is due to the fact that the local oscillator is one IF away from the received frequency and the image frequency is two IFs away, making it easier to filter. Because of this, low intermediate frequencies with high RF frequencies can result in failure to meet specifications.

An example is provided illustrating the local oscillator radiation problem and its treatment.

Example 4-15:

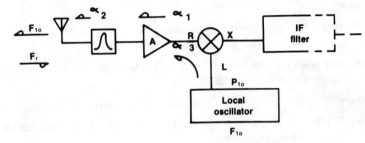

where

    F_r is the receive frequency

    F_{lo} is the local oscillator frequency

    IF is the intermediate frequency

    α_1 is the reverse gain of the preamplifier

    α_2 is preselector attenuation at F_{lo} when the receiver is tuned to F_r

    α_3 is the mixer isolation (L to R port)

The local oscillator signal power level at the antenna connector is

$$P_{lo} + \alpha_3 + \alpha_1 + \alpha_2 = P_{lo} \ (ant)$$

Illustration:

    $\alpha_3 = -20$ to -30 dB

    $\alpha_1 = -20$ dB (measured or from data sheet)

    α_2 is from the preselector curve at F_{lo} when tuned to F_r

    $P_{lo} = 10$ dBm

    $P_{lo(ant)} = -10 - (20$ to $30) - 20 + \alpha_2 = -30$ to -40 dBm $+ \alpha_2$

Find the necessary preselector attenuation required to reduce the local oscillator signal level at the antenna to -110 dBm.

Solution:

    $\alpha_2 = -110 + (30$ to $40) = 70$ to 80 dB

Such a high value of preselector attenuation demands that the local oscillator frequency be separated from the tuned frequency by an intermediate frequency that is sufficiently high to ensure the necessary attenuation is secured.

4.27 PREDICTING SPURIOUS PRODUCTS

The spurious performance of paper designs are easily verified through the use of computers.

The mixer equation was given as

$$F_{if} = |\ NF_r \pm MF_{lo}\ | \tag{4-60}$$

where

F_{if} is the intermediate frequency
F_r is the receive frequency
F_{lo} is the local oscillator frequency
M and N are integers

Since mixer generated spurious products must have an input stimulus, F_r is made the variable, F_{if} is fixed by design, and F_{lo} is solved where M and N are equal to 1.

Then

Given F_{rmax} and F_{rmin}
Choose F_{if} and F_{inc} where F_{inc} is an incremental change of F_r

Compute F_{lo} for $M = N = 1$ for every value of F_r ranging from F_{rmin} to F_{rmax} using equation (4-60) and the appropriate sign (where $F_r = F_{rmin} + F_{inc}) < F_{rmax}$.

Solve for F_s where:

$$F_s = \frac{F_{if} \pm MF_{lo}}{N} \tag{4-61}$$

and M and N take on all values from 9 to −9 in all combinations. This may be accomplished in two loops. For example, let $M = 9$ and let N run from 9 to −9 then decrement M to 8 and run N from 9 to −9, *et cetera*.

Compute DELF = $|\ F_r - F_s\ |$
Compute ORD = $|\ M\ | + |\ N\ |$
Print
$F_r, F_{lo}, F_s, DELF, M, N, ORD$

This represents a minimal program which requires reference to a mixer table. By including a look-up table in the program, as well as the preselector attenuation characteristics, all necessary information will be included. The inclusion of the preselector characteristics lets the designer explore performance against out of band signals. These computer print-outs provide a tremendous insight to spurious performance and provide the designer, management, and customer with reasonable confidence regarding performance.

A program in BASIC which is suitable for spurious prediction is shown in Table 4-7. This program has the following features:

Interactive
Prompting
Built in preselection options

Outputs

Receive frequency (F_r)
Local oscillator frequency (F_{lo})
Spur identity

harmonic of F_r (to 7th)
harmonic of F_{lo} (to 8th)

Mixer spur level up to 15th order
Total preselector attenuation at the spurious frequency
Total spurious level in dB
Spurious frequency

The program also features a search floor which is selected by the user and excludes all spurious responses below that floor.

Table 4-7.

Spurious Response Program for the Prediction
of Receiver Spurious Responses

```
10 PRINT "SPUR SEARCH PROGRAM"
20 PRINT "RF INPUT -10DBM,LO 17
   DBM"
30 PRINT "FROM"
40 PRINT "   FS=(FIF-M*FLO)/N"
50 PRINT "WHERE"
60 PRINT "   FS=SPUR FREQUENCY"
70 PRINT "   FIF=INTERMEDIATE F
   REQUENCY"
80 PRINT "   FLO=LOCAL OSC FREQ
   "
90 PRINT "   M&N ARE INTEGERS O
   F BOTH SIGNS"
100 PRINT "COMPUTES UP TO 15TH O
    RDER"
110 PRINT
120 PRINT "DEFINITIONS"
130 PRINT "   ORDER=ORD=ABS(M+N)"
140 PRINT "   FR=TUNED FREQUENCY"
150 PRINT "   FMIN=MIN LIMIT OF F
    R"
160 PRINT "   FMAX=MAX LIMIT OF F
    R"
170 PRINT
180 SHORT L,X
190 DISP "ENTER FMIN,FMAX,IF,INC
    "
```

```
200 INPUT A,B,C,D
210 PRINT "FMIN=";A,"FMAX=";B,"I
    F=";C
220 DISP "SELECT OPTION (1) IF=F
    LO-FR, (2) IF=FR-FLO, (3) IF
    =FR+FLO"
230 INPUT F
240 PRINT "OPTION=";F
250 DISP "ENTER SPUR LEVEL (DB)"
260 INPUT A1
270 PRINT "SPUR FLOOR=";A1,"DB"
280 J2=0
290 DISP "LOW PASS FILTER (Y)ES,
     (N)O ?"
300 INPUT B$
310 IF B$="Y" THEN 320 ELSE 380
320 DISP "ENTER CUT OFF,N,RIPPLE
     (DB),ULT ATTN (DB)"
330 INPUT B2,C2,D2,E2
340 PRINT "LOW PASS FILTER CUT O
    FF=";B2
350 PRINT "NUMBER OF ELEMENTS ="
    ;C2
360 PRINT "RIPPLE=";D2;"DB"
370 PRINT "ULTIMATE ATTN=";E2;"D
    B"
380 DISP "CHOOSE (FIXED) OR (TUN
    ED) FILTER"
390 INPUT A$
400 IF A$="FIXED" THEN 420
410 IF A$="TUNED" THEN 460
420 DISP "CHEBISHEV FILTER (FIXE
    D TUNED) ENTER FMIN,FMAX,N,R
    IPPLE(DB),ULTIMATE ATTN(DB)"
430 INPUT A3,A2,N1,R1,R2
440 PRINT "FIXED TUNED CHEBYSHEV
     FILTER FMIN=";A3;"FMAX=";A2
    ;"N=";N1;"RIPPLE=";R1;"ULT A
    TN=";R2
450 GOTO 490
460 DISP "ENTER QL AND NUMBER OF
     CRITICALLY COUPLED TRANSFOR
    MERS AND THE ULTIMATE ATTENU
    ATION"
470 INPUT Q1,N1,R4
```

```
480  PRINT "TUNABLE FILTER USING"
     ;N1;"TRANSFORMERS WITH A LOA
     DED Q OF";Q1;"AND ULT ATTN O
     F";R4
490  FOR A=A TO B STEP D
500  IF F=1 THEN L=C+A
510  IF F=2 THEN L=A-C
520  IF F=3 THEN L=C-A
530  PRINT "*************************
     ***********"
540  PRINT "   TUNED FREQ=";A,"FLO
     =";L
550  PRINT "FSPUR        M   N  -DB
      FLT   TOT"
560  FOR M=-8 TO 8
570  FOR N=1 TO 7
580  X=INT(ABS(C+M*L)/N*10+.5)/10
590  GOTO 650
600  INTEGER W,X3,T1
610  PRINT TAB(1);X;TAB(10);M;TAB
     (14);N;TAB(18);Q;TAB(22);X3;
     TAB(27);T1
620  NEXT N
630  NEXT M
640  NEXT A
650  U=ABS(M)
660  V=ABS(N)
670  FOR J=0 TO 7
680  FOR K=0 TO 8
690  READ W(J,K)
700  DATA 0,27,31,36,47,36,51,37,
     63,23,0,39,11,46,14,62,19,53
     ,86,75,84,75,86,74,87,74,84
710  DATA 87,77,87,78,90,75,85,77
     ,88,90,90,90,90,90,90,90,90,
     90,90,90,90,90,90,90,90,90,9
     0
720  DATA 90,90,90,90,90,90,90,90
     ,90,90,90,990,90,90,90,90,90
     ,90
730  NEXT K
740  NEXT J
750  Q=W(V,U)
760  RESTORE
770  IF Q>A1 THEN 620
780  IF B$="Y" THEN 790 ELSE 880
```

```
790  IF X>B2 THEN 820 ELSE 800
800  J2=0
810  GOTO 880
820  F2=ABS(((B2*.00001/X-X)/B2)
830  G2=C2*LOG(F2+(F2^2-1)^.5)
840  H2=(EXP(G2)+EXP(-G2))/2
850  I2=10^(D2/10)-1
860  J2=10*LGT(1+I2*H2^2*F2)
870  IF J2>E2 THEN J2=E2
880  IF A$="FIXED" THEN 900
890  GOTO 1040
900  IF X<A3 THEN 930
910  IF X>A2 THEN 930
920  GOTO 1000
930  X2=ABS((A2*A3/X-X)/(A2-A3))
940  Y=N1*LOG(X2+(X2^2-1)^.5)
950  C1=(EXP(Y)+EXP(-Y))/2
960  R=10^(R1/10)-1
970  X3=10*LGT(1+R*C1^2*X2)
980  IF X3>R2 THEN X3=R2
990  GOTO 1010
1000 X3=0
1010 T1=X3+Q+J2
1020 IF T1>A1 THEN 620
1030 GOTO 610
1040 X3=20*LGT((1+(Q1*2*(A-X)/A)
     ^4/4)^(N1/2))
1050 IF X3>R4 THEN X3=R4
1060 T1=X3+Q+J2
1070 IF T1>A1 THEN 620
1080 GOTO 610
1090 END
```

4.28 THE MIXER SPUR CHART

Several manufacturers in the mixer field have published charts describing mixer performance. Much of the data is computer generated in lieu of actual measurement. This can be appreciated when a measurement approach is attempted. Although direct measurement provides a realistic overview of mixer performance, particularly because such performance is governed by mixer termination, this approach is seldom used for economic reasons. The data provided is useful; however, it should be realized that these data are made under conditions which may vary drastically from those of the design to be analyzed. The designer must allow some margin for such variance.

Mixer spur tables assume ideal broadband terminations of usually 50 Ω, since it is impossible to cover all other possibilities and because the 50 Ω broadband termination offers the best overall performance. An example is shown in Table 4-8. The data contained can be broken down into three distinct groups. These are:

harmonics of F_r
harmonics of F_{lo}
spurious products due to mixing

The table includes values of N from 0 to 7 and M of 0 to 8. Thus, the highest order spurious term included in the table is 15. The order of any spurious term is $|M| + |N|$. For example, the fourth harmonic of F_r and the fifth harmonic of F_{lo} will result in a spurious product whose predicted magnitude is 76 dB below the desired output (where $M = N = 1$ (for a mixer signal input level of 0 dBm and a local oscillator level of +7 dBm). The order of this spur is 4+5=9. Note the effect of mixer input signal level. As the magnitude of F_r is decreased, spurious performance improves. Also, as the local oscillator drive is increased, spurious performance improves. Therefore, for optimum spurious performance from a mixer, operate it at low input signal levels, with a high local oscillator drive level.

Harmonics of F_r go straight through a mixer without benefit of conversion by the local oscillator. Where the harmonic of the input signal F_r is $1 = N$, this level is the mixer isolation or through-put and appears on most data sheets. Where the harmonic numbers are greater than one, the through-put suffers increased attenuation. This fact is important to up converter designs where it is possible for a harmonic of F_r to be equal to the first intermediate frequency.

For example, the first IF is selected to 610 MHz. The input signal F_r ranges from 50 to 250 MHz. By dividing 610 MHz by increasing interger values; note the values which fall into the range of F_r where $50 < F_r < 250$.

We find

	−dB
610/1 = 610 MHz	
610/2 = 305	
610/3 = 203.33	51
610/4 = 152.5	80
610/5 = 122	72
610/6 = 101.66	90

et cetera (where $F_r = -10$ dBm and $F_{lo} = 7$ dBm)

Such performance would not be acceptable in any designs but low grade. The poorest acceptable spurious performance level is generally 60 dB below the desired level (and 80 dB below the desired level for a quality system). The only option here is to raise the IF so that $3F_r$ or $N=3$ is excluded. Where cost is a factor, customers often will waive specifications for a few such isolated spurious responses.

Similarly, F_{lo} and its harmonic will appear at a mixer's output without regard to the input frequency F_r. Where the harmonic of $F_{lo} = 1$ or $M = 1$, this value is called the F_{lo} to IF mixer isolation and appears in the mixer data sheet. This value is indicative of the balance of a mixer.

Where the harmonic number of F_{lo} is greater than one, these will appear at the mixer's IF output port with increased attenuation. It is undesirable to have harmonics of F_{lo} falling within the IF. This can cause heterodyne whistles, birdies, or desensitization of a receiving system at these frequencies.

Example 4-15:

$50 < F_r < 250$ MHz

IF = 610 MHz

We find

$$F_{lo} = \text{IF} - F_r$$
$$= 610 - (50 < F_r < 250)$$

then

$$560 > F_{lo} > 360$$

solving for

$KF_{lo} = \text{IF}$ or $K = \text{IF}/F_{lo}$, where K is an integer

$$K = \frac{610}{560 > F_{lo} > 360}$$

We see for this case there are no local oscillator harmonic problems. Here it is avoided by parameter choice using up conversion. Should the IF have been 403 MHz, then

$$353 > F_{lo} > 153$$
$$K (353 > F_{lo} > 153) = 403$$

the results would have been:

K	IF$/K$	dB	
1	403		(out of F_r range)
2	201.5	45	
3	134.33	52	
4	100.75	63	
5	80.6	45	

where $F_{lo} = 7$ dBm.

This represents poor performance and the designer should reconsider the IF selection. Note that increasing the level of F_{lo}, while reducing spurious product levels, where $M > 1$ and $N > 1$, increases the F_{lo} harmonic problem. For more information on IF selection see Section 4-5.

4.28.1   Spur Chart Limitations

A spur chart assumes ideal broadband terminations seldom realized in practice. Further, it is derived at a single value of intermediate frequency. The chart rarely totally applies to the design for which it is used as an analysis tool. It does serve as a valuable aid, a step better than a mixer spur graph, but the ultimate test is an actual receiver measurement. The designer should allow a design margin when using spur chart values in analysis.

4.28.2   Spur Chart Generation by Measurement and Spur Identification

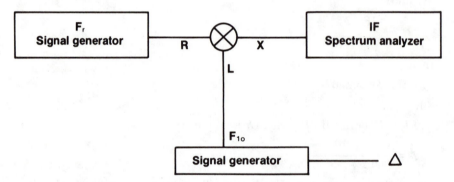

Fig. 4-31. Test setup for spur chart measurement and spur identification.

This technique, though tedious, is easily implemented as shown in Fig. (4-31). The IF and a value of F_r are selected. The spectrum analyzer serves as the IF filter and provides spur product magnitude and identification. A value of F_{lo} is computed where:

$$F_{lo} = |\text{ IF} \pm F_r| \text{ (sign as applies)}$$

The mixer under test is provided the F_{lo} input signal at the desired level. This remains fixed for the spur search duration.

The spectrum analyzer is tuned to the IF and remains fixed. The signal generator providing the F_r signal is set to the desired test level and the spectrum analyzer is calibrated to a reference level on the signal where the relationship IF $= |F_{lo} \pm F_r|$ is satisfied. The frequency F_r is then varied over the spur search frequency range while observing the spectrum analyzer for any other responses. When one is found, note the magnitude relative to the desired signal.

The spur may be identified by incrementing either F_r or F_{lo} first, by a known amount, noting the frequency shift of the spur and then repeating the process for other F_r or F_{lo} values.

Example 4-17:

A spur is found and its magnitude is 62 dB below the desired reference output. The F_r source is shifted by a small convenient amount Δ MHz and the spectrum analyzer display of the spur is seen to shift 3Δ. This is a third harmonic of F_r, so $N = 3$.

Shifting the frequency of F_{lo} by Δ'. Therefore, this is a fourth harmonic F_{lo} and $m = 4$. The spur is then identified as resulting from a third harmonic of F_r and a fourth harmonic of F_{lo}. The order is seventh. The −62 dB level is entered in the $3F_r$ by $4F_{lo}$ location of the chart. This process is repeated until all slots are filled. Where no spurs are found, enter a level equal to or greater than the search threshold or sensitivity. In addition to spur chart generation, this process also serves to identify troublesome spurs on a systems level.

4.29 LOCAL OSCILLATOR SPURIOUS PRODUCTS

Local oscillators are seldom totally pure and at the least contain some LO harmonic signals. Synthesized sources will contain clock spurs, which are clock fundamental and harmonic frequencies, located on both sides of the main signal, in addition to the main signal harmonics. Where mixers are employed in the output of the local oscillator chain, mixer generated spurs can be added to those previously mentioned resulting in further spectral contamination. The result of all this is a worsening of the receiving system's spurious performance, because of LO impurity.

The LO signal, spurious input terms, and their relative magnitudes presented to a mixer, together with a pure RF input signal, will produce an IF output which is a replica of the LO signal. However, the IF output is scaled to the magnitude of the RF input signal less conversion loss of the mixer. The replication will be reversed spectrally if the LO is below the received signal.

An illustration of this effect is shown in Fig. (4-32).

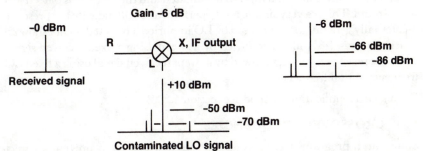

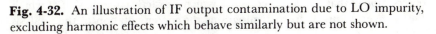

Fig. 4-32. An illustration of IF output contamination due to LO impurity, excluding harmonic effects which behave similarly but are not shown.

Harmonics of the LO will cause spurious mixer outputs scaled to the main LO signal, as before. If they are particularly high they can cause serious additional spurious outputs. Consider as a limit the configuration of Fig. (4-33). Here the LO source contains an unwanted spur which is equal to that of the desired signal. The result is a set of spurious outputs which satisfy the relationship

$$\text{IF} - mF_{lo} \pm nF_r \pm pF_{lo} \text{ (spur)}$$

where m, n, and p are integers

It is not beneficial to pursue the solution of this relationship because of the three-dimensional behavior. Spur tables are available only up to two dimensions making it impossible to make any useful spurious predictions. The best approach is direct measurement.

A partial solution can be had by first letting $p = 0$ and solving for F_r for all values of m and n which is the ideal case, and then letting $m = 0$ and solve again for all m and p values.

The use of impure LO sources should be avoided unless the design is of very low performance intentionally.

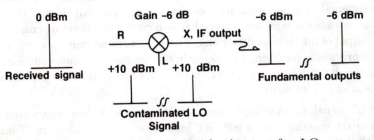

Fig. 4-33. An illustration of mixer output for the case of an LO spur equal to the LO signal itself.

Local oscillator spurs can cause receiver outputs as the receiver is tuned, even though the receiver antenna input is terminated and the receiver is placed in a screen room. This is a very simple test to perform and will indicate the presence of internally generated spurs caused by LO impurity. This test does not show all spurious terms. It should be followed by a sweep using a fixed received signal while the receiver is tuned, followed by a swept received signal using a fixed LO frequency for all frequencies.

For a spur identification method see section 4.28.2

4.30 CROSSOVER FREQUENCIES

Spur search programs have a resolution limit which often prevent a direct observation of that spurious frequency, which is exactly equal to that at which the receiver is tuned. In such programs the incrementation of the analysis results in offsets between the receive and spurious frequencies. While it is possible to predict such points of frequency equality by extrapolation of computer data it becomes a tedious task.

Spurious frequencies are migratory with the exception of harmonics of the local oscillator, harmonics of the received frequency, and subharmonics of the intermediate frequency. As the receiver frequency is tuned, all other spurious terms will change their spectral positions. Some spurious terms will diverge from the receive frequency, others will converge, while some will actually cross through the receive frequency. The latter are called crossover spurious frequencies. They are important in that the point where the spurious and receive frequencies are equal or within the IF passband, the receiver will see two signals simultaneously. The result of this will be heterodynes (or beats), desensitization, or both, depending upon the type of receive system. It is desirable to find these frequencies and determine their magnitude since no preselection filtering will be effective in their removal.

A simple relationship determines the crossover spurious frequencies.

Case 1 Difference mode.

$$F_{lo} = \left| F_{if} - F_r \right| \tag{4-62}$$

and

$$F_{if} = \left| mF_{lo} + nF_s \right| \tag{4-63}$$

substituting (4-62) into (4-63)

$$F_{if} = \left| mF_{if} - mF_r + nF_s \right| \tag{4-64}$$

but

$F_r = F_s$ to be a crossover term.

then

$$F_{if} (1-m) = F_s (n-m) \tag{4-65}$$

$$F_s = \left| F_{if} \frac{(1-m)}{(n-m)} \right| \tag{4-66}$$

Case 2 Sum mode

$$F_{lo} = \left| F_{if} + F_r \right| \tag{4-67}$$

and

$$F_{if} = \left| mF_{lo} + nF_s \right| \tag{4-68}$$

substituting (4-67) into (4-68)

$$F_{if} = \left| mF_{if} + mF_r + nF_s \right| \tag{4-69}$$

but

$$F_r = F_s$$

then

$$F_{if}(1-m) = F_s(m+n) \tag{4-70}$$

$$\therefore F_s = \left| F_{if} \frac{(1-m)}{(m+n)} \right| \tag{4-71}$$

Where

F_r is the desired receive frequency
F_s is the spurious frequency
F_{lo} is the local oscillator frequency
F_{if} is the intermediate frequency
m and n are integers of either sign
(a useful range of m and n is 9, to −9 or 18th order maximum)

The solution of these simple equations is solvable on hand-held calculators but is somewhat slow and should be solved by desk top or better machines.

4.31 COMPUTING NOISE FIGURE GIVEN TSS

From the definition of TSS as 8 dB output S/N, noise figure may be computed from:

$$NF = TSS - 8 - kTB \tag{4-72}$$

Example 4-18:

Given

TSS = −90 dBm

kTB = −114 dBm, (B = 1 MHz)

Solution:
NF = −90 − 8 − (114) = 16 dB

REFERENCES

[1] Frutiger, "Noise in FM Receivers with Negative Feedback," *IEEE*, Vol. 54, Nov. 1966.

[2] Schwartz, M., *Information Transmission and Noise.* New York: McGraw-Hill, 1959.

[3] Emde, Jahnke, and Losh, *Tables of Higher Functions,* 6th ed. New York: McGraw-Hill, 1960.

[4] McVay, Franz C., "Don't Guess the Spurious Level," *Electronic Design,* Vol. 3, Feb. 1, 1967.

[5] Norton, David E., "The Cascading of High Dynamic Range Amplifiers," Waltham, MA: Anzac Electronic.

[6] Goldberg, Harold, "Predict Intermodulation Distortion", *Electronic Design*, Vol., 10, May 10, 1970.

Table 4-8
(Courtesy Watkins-Johnson Company)

SINGLE-TONE INTERMODULATION DISTORTION

Each cell is shown as: (fR @ 0 dBm: A B C) / (fR @ 10 dBm: A B C)

HARMONICS OF fR \ HARMONICS OF fL	0	1	2	3	4	5	6	7	8
7	79>99>99 / >90>90>90	69 79>99 / >90>90>90	80>99>99 / >90>90>90	74 78>99 / >90>90>90	83>99>99 / >90>90 90	63 78>99 / 87>90>90	78>99>99 / >90>90>90	60 81>99 / >90>90>90	71 99>99 / >90>90>90
6	90>99>99 / >90>90>90	86>99>99 / >90>90>90	91>99>99 / >90>90>90	91>99 97 / >90>90>90	90>99>99 / >90>90>90	84>99>99 / >90>90>90	93>99>99 / >90>90>90	84>99>99 / >90>90>90	88>99 98 / >90>90>90
5	72 93>99 / >90>90>90	70 73 96 / 80>90>90	71 87>99 / >90>90>90	52 72 95 / 71>90>90	77 88>99 / >90>90>90	46 66>99 / 68>90>90	75 85>99 / >90>90>90	45 64 90 / 65>90>90	73 82>99 / 88>90>90
4	80 96 88 / 86>90>90	79 80 91 / >90>90>90	82 96>99 / 86>90>90	77 80 92 / 88>90>90	82 95 90 / 88>90>90	76 82 95 / 85>90>90	77 98 87 / 86>90>90	72 78 94 / 85>90>90	77 90 87 / >90>90>90
3	51 63 81 / 67 87>90	49 58 73 / 64 77>90	53 65 85 / 69 87>90	51 60 69 / 50 78>90	55 65 85 / 77>90>90	48 55 68 / 47 75>90	54 64 85 / 74 85>90	53 54 64 / 44 77 89	58 66 87 / 74 88>90
2	69 68 64 / 73 86 73	72 67 71 / 73 75 83	79 76 62 / 74 84 75	67 67 70 / 70 75 79	75 80 63 / 71 86 80	66 66 70 / 64 74 80	77 82 61 / 69 87 77	68 66 62 / 64 74 82	75 83 64 / 69 84 79
1	25 25 24 / 24 23 24	0 0 0 / 0 0 0	39 39 35 / 35 39 34	13 11 11 / 13 11 11	45 50 42 / 40 46 42	22 16 19 / 24 14 18	54 59 50 / 45 62 49	37 19 39 / 28 19 37	59 59 49 / 49 53 49
0	A B C	36 39 29 / 26 27 18	45 42 20 / 35 31 10	52 46 32 / 39 36 23	63 58 24 / 50 47 14	45 37 29 / 41 36 19	60 65 27 / 53 51 17	71 49 30 / 49 37 21	64 75 29 / 51 63 19

Legend:

| fR @ 0 dBm |
| fR @ 10 dBm |

fR = 49 MHz
fL = 50 MHz

HARMONICS OF fL

A: (M1)	B: (M1D/M9BC)	C: (M1E/M9E)
0.2 – 500 MHz	0.5 – 500 MHz	1.0 – 400 MHz
CLASS I MIXER	CLASS II	CLASS III
LO = +7 dBm	(TYPE 2)	LO = +27 dBm
	LO = +17 dBm	

5

COMPONENTS

Vital to the characterization of a receiving system and the prediction of its performance is a knowledge of the components used and their capabilities. This chapter addresses this area in a practical sense. The goal is not to design components but to assist the designer in the selection of a component and assign achievable performance values to that component.

Several computer programs are included which allow the user to determine the theoretical performance of some of the more popular forms. It is strongly recommended that the reader develop a library of components from manufacturers specializing in their design. This is invaluable for reference purposes and future procurement.

5.1 FILTERS

The filtering problem is solvable through the use of several popular filter types. Included are:

LC

Crystal

Helical

Tubular

Cavity

It is not the intent of this text to examine this subject in any detail except to point out the limits of the operating frequency of each type together with the range of achievable bandwidths. The data is necessarily approximate but serves as a guide for the designer. Where requirements are out of these bounds it is best to contact a supplier to see if the art has been extended and the design is feasible.

The LC filter is capable of operating over a range of roughly 100 Hz to several GHz with bandwidths ranging from approximately 1/2 to 130%. The particular range of realizable bandwidth *versus* frequency is shown in Fig. (5-1).

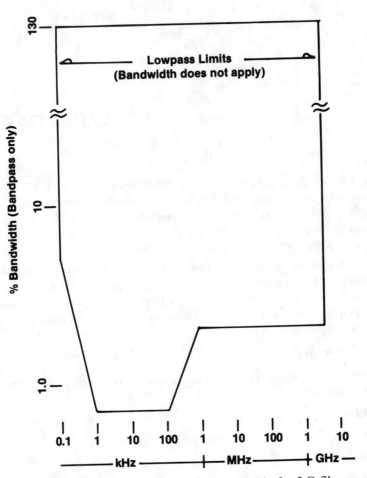

Fig. 5-1. Approximate range of bandwidths available for LC filters *versus* frequency (©1982, K&L Microwave, Inc.)

Helical resonator filters are easily realized over a frequency range of approximately 20 MHz to 1 GHz, with bandwidths of 0.2 to 3, or more (typically 15%).

Tubular filters, while bulky, are useful in some applications. This class of filter is available in either lowpass or bandpass designs. The lowpass filter is available from 10 MHz to 18 GHz and the bandpass type is available in the 30 MHz to 12.4 GHz range. Bandpass filter bandwidths are in the 1 to 80% area.

Typical helical and tubular filter characteristics are shown in Fig. (5-2).

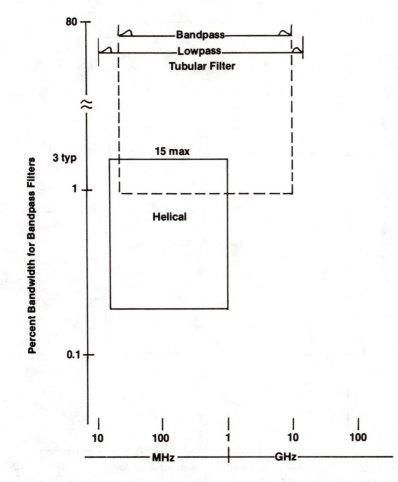

Fig. 5-2. Percent bandwidth and frequency range for tubular and helical bandpass filters. For the lowpass tubular filters, the percent bandwidth does not apply (©1982, K&L Microwave, Inc. and an April 1976 TelonicAltair catalog)

Cavity filters, which are popular for the first conversion filter in up conversion receiver designs, are capable of narrow bandwidths (typically 0.1 to 3.5%) in the frequency range of 30 MHz to over 40 GHz. Wideband filters of this class are available with bandwidths of 5 to 50% at frequencies ranging from 200 MHz to over 18 GHz. Fig. (5-3) illustrates these characteristics.

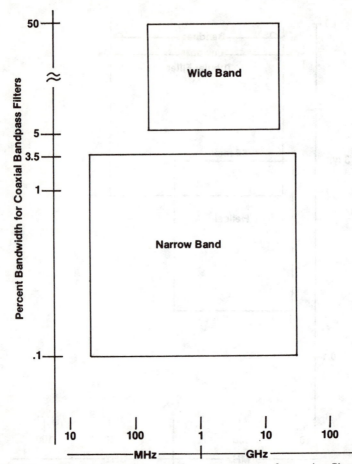

Fig. 5-3. Percent bandwidth and frequency range for cavity filters (©1982, K&L Microwave, Inc.)

The characteristics of crystal filters are discussed in section 5.1.4. Several other filters are available and are worth mentioning. This includes the interdigital filter, which, when using strip line techniques and the air dielectric, a 3 to 30% bandwidth is achievable over the frequency range of 1 to 5 GHz.

A second important filter class is the comb-line, which is capable of 1 to 15% bandwidths over the frequency range of 1.3, to greater than 20 GHz with the air dielectric.

Both of these filters are made in other forms and utilize dielectrics other than air. Generally these filters are relatively high frequency types. The size of the filter is a function of the number of elements as well as the type of dielectric. It is suggested that suppliers be contacted for details.

One of the most popular filter types is the Chebyshev because of its steep selectivity. This is achieved at a sacrifice of phase linearity. These filters have an inherent ripple, which results in distortion of certain waveforms. Where this is a concern, other filter types should be considered. The Bessel and the Gaussian filter have a good phase linearity characteristic but with a sacrifice in skirt selectivity. Here the response is parabolic. The Gaussian filter has a rounded group delay characteristic where the Bessel filter has a flat group delay.

In some applications it may be necessary to match the phase of filters. Where this is the case, such performance is typified by Fig. (5-4). This is a somewhat generalized illustration and it must be kept in mind that such matching is more difficult in complex designs.

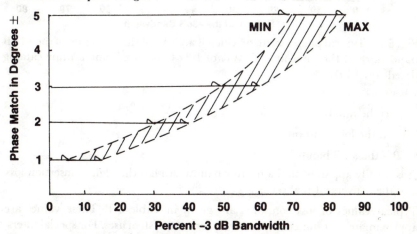

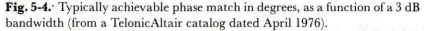

Fig. 5-4. Typically achievable phase match in degrees, as a function of a 3 dB bandwidth (from a TelonicAltair catalog dated April 1976).

Phase linearity is often important. This can be approximated by reference to filter charts for particular types. In general, for a 1.3/1 VSWR condition, the phase linearity illustrated in Fig. (5-5) is achievable. Again reference should be made to filter curves for specific cases.

5.1.1 Filter Insertion Loss, Bandpass Case

While there are equations for filter insertion losses, the most practical approach, except for special cases, is the use of the loss constant (L_k). For any filter the insertion loss (I_l) is simply:

$$I_l = \frac{(N + 0.5)}{\% B} L_k , \ dB \tag{5-1}$$

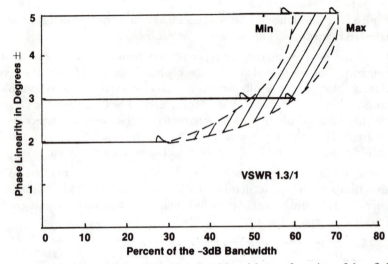

Fig. 5-5. Typical phase linearity which is achievable as a function of the –3 dB bandwidth of the filter at a VSWR of 1.3/1 (from a TelonicAltair catalog dated April 1976).

where

   N is the number of sections

   L_k is the loss constant

   B is the 3 dB bandwidth

It is readily apparent that a narrow bandwidth has the highest insertion loss, and the converse it is also true.

Typical values of loss constants are given in Table 5-1. These values are approximate, but sufficiently accurate for most system uses. For special filters, or where fractions of a dB are of concern, filter specialists should be consulted.

5.1.2 Filter Insertion Loss: Lowpass and Highpass Cases

The insertion loss (I_l) equation for the lowpass and highpass cases is:

$$I_l = N L_k \tag{5-2}$$

where

   N is the number of sections

   L_k is the loss constant

Table 5-2 lists some typical values for various filters. These values serve as a guide in system design. Where the design is critical, or special filters are used, filter experts should be consulted.

Table 5-1.

Approximate Loss Constants for Bandpass Filters

Type & Ripple	Frequency (MHz)										
	30 50	50 65	65 100	100 400	400 600	600 900	900 1300	1300 1800	1800 3000	3000 10000	10000 12000
Tubular .05 dB											
.25″ dia.				5	4	4	3.5	3.5	2.5	2	
.375″ dia.				4	3	2.5	2	2	1.6		
.5″ dia.		4	3.5	3	2	2	1.4	1.4	1.1		
.75″ dia.	3.5	2.5	2.2	2	1.4	1.4	1.2	1.2			
1.25″ dia.	2.5	2.2	1.8	1.6	1.2	1.2					
LC 0.1 dB			5 to 6								
Micro miniature 0.1 dB ripple	6.8			6.8 to 5.7	4.9 to 4	4 to 3.25	3.25	3.0			
Cavity	1.7	1.6	1.5	1.4	1 to 3	1 to .25	.35 to .25	.3 to .22	.3 to .2	.45	.35
Miniature Cavity				3 to 2.5		2					

Table 5-2.

Approximate Loss Constants for High- and Lowpass Filters

Type & Ripple	Frequency (MHz)									
	10 25	25 50	50 100	100 250	250 500	500 1000	1000 2000	2000 4000	4000 6000	6000 18000
Tubular .05 dB										
.25″ dia.				.3	.25	.25	.2	.18	.1	.1
.375″ dia.				.25	.2	.2	.18	.16		
.5″ dia.		.2	.18	.16	.16	.13	.11	.1		
.75″ dia.		.18	.14	.13	.13	.12	.11			
1.25″ dia.	.14	.12	.09	.08	.08	.07				
LC .01 dB		.1 to 1.4								
Micro min. LC .1 dB ripple					.25	.15 to .2	.1 to .15	.1 to .15	.1 to .15	

5.1.3   Varactor Tuned Filters

The varactor tuned or voltage tuned filter is an LC filter, where C is the variable supplied by a varactor tuning diode.

These filters are usually two or four pole designs, or cascades of two pole filters. Tracking becomes a problem as more poles are used. These filters are most often critically coupled tuned circuits and are available in constant Q, or constant bandwidth designs.

Typical performance from a two pole design is shown in Table 5-3. These filters are used with a tuning linearizer for remote tuning purposes. In manually tuned applications, a simple potentiometer serves as the control element.

Table 5-3.

Typical Performance of Two-Pole Varactor-Tuned Filters

Available frequency range	10 MHz to 2 GHz
Tuning range	1 octave
Bandwidth	4 to 12%
Insertion loss	≈2 dB
Shape factor 3/30 dB	≈ 5
Input signal power	2 watts
Tuning time	microseconds
Tuning voltage	0 to 20 or 60 volts
Size	1″ by 1″ by 0.5″
Weight	1.5 oz

The varactor tuned filter is often designed in house because of its relative simplicity. The design usually takes on the form of two tuned circuits coupled together. Where this configuration does not provide the necessary selectivity, two or more (usually not more than three) tuned circuit pairs are placed in series, with amplifier isolation between each pair.

There are several popular configurations in use. They are shown in Figs. (5-6) through (5-8).

Coupling into the filter from the input to the primary tuned circuit can take on any of the impedance transformation configurations. This applies equally well to the output tuned circuit coupling. Of particular interest is tap or link coupling. If this is accomplished through a series inductance, the coupling loss of the filter can be made nearly flat. This is illustrated in Figs. (5-9) and (5-10).

While the previous discussion has ignored the varactor diode, or tuning diode, there is little significant alteration of the filter, except for a degradation of Q power handling, and intermodulation distortion of the end configuration.

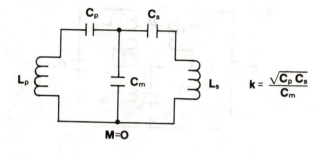

$$k = \frac{\sqrt{C_p C_s}}{C_m}$$

A. M=0

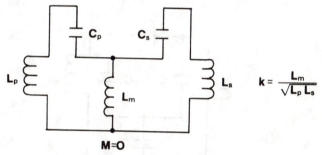

$$k = \frac{L_m}{\sqrt{L_p L_s}}$$

B. M=0

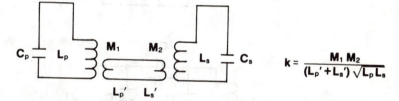

$$k = \frac{M_1 M_2}{(L_p' + L_s') \sqrt{L_p L_s}}$$

C.

Fig. 5-6. Low impedance coupling forms of two coupled tuned circuits.

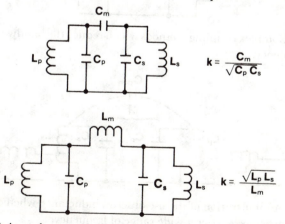

$$k = \frac{C_m}{\sqrt{C_p C_s}}$$

$$k = \frac{\sqrt{L_p L_s}}{L_m}$$

Fig. 5-7. High impedance coupling examples of two coupled tuned circuits.

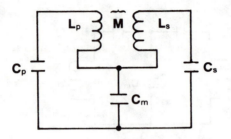

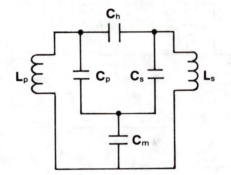

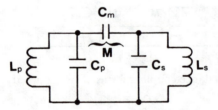

Fig. 5-8. Complex coupling methods which reduce the bandwidth dependency on frequency.

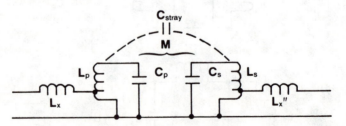

Fig. 5-9. An illustration of the use of series inductance when coupling to the filter, to flatten losses over a wide range of frequency.

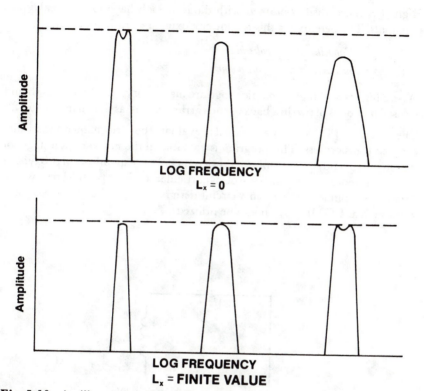

Fig. 5-10. An illustration of the effect of L_x on loss as a function of frequency.

The tuning diode is one where the diode capacity, as a function of reverse bias, is controlled and enhanced. The operating range of such a diode is from cutoff to below reverse breakdown. In its equivalent form it consists of a series parallel arrangement of R, C and L as shown in Fig. (5-11).

Data sheets will generally specify the values of the parameters of Fig. (5-11). The reverse bias is usually given at four or six volts in a circuit application at 1 MHz. Diode capacitance will be given as C_{tv} where v is the bias potential. This value gives the designer a measure of the device's capacitance for comparative use.

The Q of the diode is often neglected by the designer resulting in selectivity degradation. The data sheet will often provide Q_v values (at a bias of v volts) or provide C_{tv} and R_{sv} at some frequency f such as 50 MHz. Then:

$$Q_v = \frac{1}{2\pi f R_{sv} C_{tv}} \text{ relates these parameters} \tag{5-3}$$

High Q is more easily obtained with diodes which have lower breakdown potentials. The penalty for this is a lower capacitance ratio given by:

$$\frac{C_{vo}}{C_{vbr}} = \frac{Capacitance\ at\ 0\ volts\ bias}{Capacitance\ at\ reverse\ breakdown} \tag{5-4}$$

Where high Q and high dynamic range are mandatory, varicap diodes should be used in groups of two in a back to back arrangement, as shown in Fig. (5-12).

The program of Table 5-4 may be used to evaluate the performance of critically coupled transformers. This program is valuable in the initial design stages of system design and serves to inform the designer of the capabilities of this type of filter. For more exact solutions of a particular design, particularly where complex coupling is utilized in a circuit design phase, a circuit analysis program such as COMPACT should be utilized.

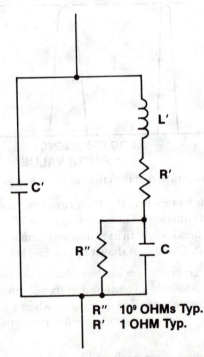

R″   10⁹ OHMs Typ.
R′   1 OHM Typ.

Fig. 5-11. Equivalent circuit of a varactor diode where C' and L' are the parasitic capacity and inductance, respectively; R' is the sum of the bulk resistances; R'' is the leakage resistance of the junction; and C is the capacitance. The latter three parameters are a function of reverse bias.

57	Equation (4-1) f_i should be f_1 to agree with definition under "where"
58	First line following Fig. 4-1 (M1) should be (M$_1$)
59	Following Fig. 4-3 M1 and M2 should be M$_1$ and M$_2$. ref. lines 2 and 5.
70	Line 5 "and" should be "in"
74	Equation (4-17) G_i should be G_1
77	Table 4-1 third column 111 should be 11.1
82	Equation (4-32) $(S/N)_{out} =$ should be opposite the denominator line. Missing (lower right) should be $\dfrac{}{(1 - e^{-(P_{c_i}/N_i)})^2}$
83	First line following Example 4-4 $(S/N)_{ou}$ should be $(S/N)_{out}$
104	Example 4-7

Receiving Systems Design
by Stephen J. Erst
© 1984 Artech House, Inc.
ERRATA

Page	Corrections
12	Fig. 2-3 (b) Add 2 diagonal lines from F − f and F + f as in (a) to the sidebands for clarity.
13	First line below Fig. 2-4 … in the carrier plus that in …
15	subscript errors top of page (P_r) and (P_t) should both be (P_e)
18	Equation (2-13) subscript A_j should be the same as that contained in the first line which follows. "The generalized solution of A_j is shown in Fig. (2-9)."
21	Equation between (2-19) and (2-20) $2\pi f$ should be $2\pi F$.
22	Equation (2-23)

Fig. 2-13

½ should be 2

36   Fig. 2-20
Table titled "DPSK encoded data to transmitter" heading
should be

     a   b   x $\bar{x}$

40   2.8.2 second paragraph first line
change "mean" to "means"

49   Equation (3-17) following "where" change definition of k as
follows:

k is Boltzmann's constant $(1.38044 \pm 0.00007) \cdot 10^{-23}$ joules/°Kelvin

50   Equation following (3-23) change as follows:

$$F_t = 1 + \frac{T_{eff_t}}{T_o} = 1 + \frac{T_{eff_1}}{T_o} + \frac{1 + \dfrac{T_{eff_2}}{T_o} - 1}{G_1}$$

51   Figure 3-5 change NF = –7.5 to NF = 7.5

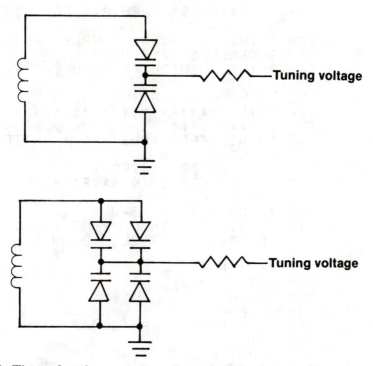

Fig. 5-12. The preferred arrangement of tuning diodes for lower distortion, higher *Q*, and power handling capability.

Table 5-4.

Computer Program for Critically Coupled Transformer Response

```
10 INTEGER X3
20 SHORT F
30 DISP "CRITICAL COUPLED TRANS
   FORMER/S RESPONSE"
40 DISP "DEFINE CENTER FREQUENC
   Y(MHZ),QL,NUMBER OF XFORMERS
   "
50 INPUT F1,Q,N
60 DISP "DEFINE FREQUENCY SWEEP
   ; START, STOP, AND INCREMENT
   (MHZ)"
70 INPUT A,B,I
80 DISP "ENTER ULTIMATE ATTENUA
   TION (DB)"
90 INPUT U
100 DISP "GRAPH? YES NO"
110 INPUT A$
```

```
120 IF A$="YES" THEN 240 ELSE 13
    0
130 PRINT "CRITICAL COUPLED TRAN
    SFORMER/S RESPONSE"
140 PRINT "CENTER FREQUENCY=";F1
    ;"MHZ"
150 PRINT "QL=";Q;"N=";N
160 PRINT "***********************
    ************"
170 PRINT "FREQ (MHZ)          ATTE
    N (DB)"
180 FOR F=A TO B STEP I
190 X3=20*LGT((1+(Q*2*(F1-F)/F1)
    ^4/4)^(N/2))
200 IF X3>U THEN X3=U
210 IF A$="NO" THEN PRINT TAB(1)
    ;F;TAB(19);X3 ELSE DRAW F,-X
    3
220 NEXT F
230 GOTO 610
240 GRAPH
250 GCLEAR
260 DISP "ENTER YAXIS MAX ATTN(D
    B)"
270 INPUT K
280 SCALE A,B,-K,0
290 DISP "ENTER XAXIS TIC MARKS
    (MHZ)"
300 INPUT M1
310 DISP "ENTER YAXIS TIC MARKS(
    DB)"
320 INPUT M2
330 XAXIS -K,M1
340 YAXIS A,M2
350 REM LABEL XAXIS
360 LDIR 90
370 FOR X=A+2*M1 TO B STEP M1
380 MOVE X,-(K*.9)
390 LABEL VAL$(X)
400 NEXT X
410 REM LABEL YAXIS
420 LDIR 0
430 FOR Y=-K TO 0 STEP M2
440 MOVE A+(B-A)/20,Y
450 LABEL VAL$(Y)
460 NEXT Y
470 MOVE A+(B-A)/5,.2*(-K)
480 LABEL "CRITICAL COUPLED"
```

```
490 MOVE A+(B-A)/5,.25*(-K)
500 LABEL "TRANSFORMER/S"
510 MOVE A+(B-A)/5,.3*(-K)
520 LABEL "QL="&VAL$(Q)
530 MOVE A+(B-A)/1.5,.3*(-K)
540 LABEL "N="&VAL$(N)
550 MOVE A+(B-A)/2.2,.97*(-K)
560 LABEL "(MHZ)"
570 MOVE A+(B-A)/7,.5*(-K)
580 LABEL "(DB)"
590 MOVE A,-K
600 GOTO 180
610 END
```

Example 5-1:

Center frequency?	90 MHz
Q_1 ?	40
No. of Transformers?	2
Sweep Start?	70 MHz
Sweep Stop?	110 MHz
Increment?	1 MHz
Ultimate Attenuation?	60 dB
Graph?	Yes

(It is suggested that a No be used first to establish the scaling from a print out table)

Y axis max attenuation?	70 dB
X axis TIC	5 MHz
Y axis TIC	10 dB

The result of these inputs is shown in Fig. 5-13. It is not required to enter the scaling such as MHz or dB. These are a part of the program scale factors.

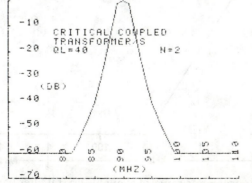

Fig. 5-13. The computer output for the graph condition for Example 5-1. This graph is valuable in the determination of attenuation capabilities for critical frequencies such as: IF rejection, IMAGE rejection, LO leakage to the antenna, *et cetera.*

5.1.4   Crystal Filter

The crystal filter is the one most widely used in IF applications because of its inherent stability, low cost and good performance. The crystal filter exhibits a typical insertion loss of 3 dB, a phase linearity of 10% over a 75% bandwidth, with shape factors of 3 to 5.5 and for special cases (with poorer phase linearity) bandwidths of 200 Hz with shape factors (3 to 60 dB) of 1.1 to 1.

A graph of typically attainable performance is shown in Fig. 5-14. Shown, are bandwidths obtainable as a function of center frequency. The graph serves as a guide to performance. Borderline cases should be verified with manufacturers, since breakthroughs are possible.

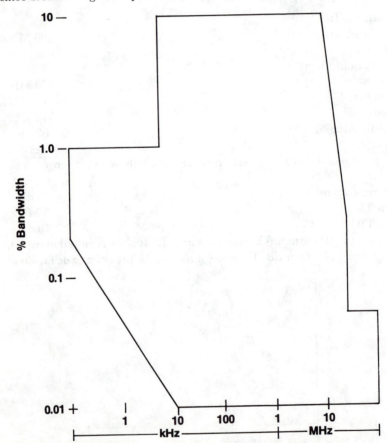

Fig. 5-14. Approximate range of available crystal filter bandwidth as a function of frequency.

Because of the very high Q of crystals, the usual filter consists of several crystals staggered in frequency throughout the bandpass (resulting in ripple). Ripple may be of concern to the designer since distortion can result, particularly with FM systems.

5.1.4.1 *Monolithic Crystal Filters*

The monolithic crystal filters are a type of crystal filter where the input and output resonators are deposited on opposite sides of a piezo electric wafer. These very small, low cost devices are capable of operation in the 5 to 350 MHz range with inductorless bandwidths ranging from 0.3 to 2% of the center frequency.

Because of their low cost and small size, monolithic crystal filters can be used to good advantage in channelized receiver front ends, as well as in intermediate frequency applications.

5.1.5 Modeling the Butterworth Filter

The Butterworth filter response is described by:

$$A\ (\ dB\ )\ =\ 10\ \log_{10}(\ 1\ +\ (\ k\ )^{2n}\ ) \tag{5-5}$$

where

$k = b_x/b_3$ for low- and bandpass filters

$k = b_3/b_x$ for highpass and notch filters

b_3 is the 3 dB bandwidth of the filter

b_x is the bandwidth at the frequency x

n is the number of resonators for band and notch filters and is the number of reactances for the high and lowpass cases

This equation accurately describes the response of any of the four filter classes. The response predicted by this equation is approached in practice, and for this reason a margin must be allowed when going from theory to practice.

When making calculations of a Butterworth response, a finite value of ultimate attenuation should be selected to truncate the calculated attenuation. Typically, a value of 60 dB is attainable without much difficulty and values to 100 dB are approached.

To assist the reader the Butterworth equation has been programmed in the BASIC language and is shown in Table 5-5. This program has a print or plot option, is prompting and interactive, requiring the operator to only answer the questions asked. Some modification may be necessary in the plot routine to accommodate the particular computer used. (This program was written for the HP 85 machine.)

Table 5-5.

Butterworth Filter Program for Low-, High-, Bandpass, and Reject Filters

```
10 SHORT F
20 DISP "BUTTERWORTH FILTER RES
   PONSE"
30 DISP "SELECT FILTER TYPE (L)
   OW,(B)AND,(H)I PASS,OR (R)EJ
   ECT"
40 INPUT B$
50 DISP "ENTER 3DB POINTS FMIN,
   FMAX"
60 INPUT F1,F2
70 DISP "ENTER N"
80 INPUT N1
90 DISP "DEFINE FREQUENCY SWEEP
   ; START, STOP, AND INCREMENT
   (MHZ)"
100 INPUT A,B,I
110 DISP "ENTER ULTIMATE ATTENUA
    TION (DB)"
120 INPUT U
130 DISP "GRAPH? YES NO"
140 INPUT A$
150 IF A$="YES" THEN 390 ELSE 16
    0
160 PRINT "BUTTERWORTH FILTER RE
    SPONSE"
170 PRINT "BAND WIDTH=";F1;"TO";
    F2;"MHZ"
180 PRINT "N=";N1
190 PRINT "*******************
    *********"
200 PRINT "FREQ (MHZ)        ATTE
    N (DB)"
210 FOR F=A TO B STEP I
220 IF B$="B" THEN GOTO 260
230 IF B$="L" THEN GOTO 280
240 IF B$="H" THEN GOTO 320
250 IF B$="R" THEN GOTO 300
260 B1=2*ABS(F-((F2-F1)/2+F1)) @
    B3=F2-F1 @ X3=10*LGT(1+(B1/
    B3)^(2*N1))
270 GOTO 340
280 B1=F @ B3=F2 @ X3=10*LGT(1+(
    B1/B3)^(2*N1))
290 GOTO 340
```

```
300 B1=2*ABS(F-((F2-F1)/2+F1)) @
    B3=F2-F1 @ X3=10*LGT(1+(B3/
    B1)^(2*N1))
310 GOTO 340
320 B1=F @ B3=F1 @ X3=10*LGT(1+(
    B3/B1)^(2*N1))
330 GOTO 340
340 IF X3>U THEN X3=U
350 X3=INT(X3*100+.5)/100
360 IF A$="NO" THEN PRINT TAB(1)
    ;F;TAB(19);X3 ELSE DRAW F,-X
    3
370 NEXT F
380 GOTO 740
390 GRAPH
400 GCLEAR
410 DISP "ENTER YAXIS MAX ATTN(D
    B)"
420 INPUT K
430 SCALE A,B,-K,0
440 DISP "ENTER XAXIS TIC MARKS
    (MHZ)"
450 INPUT M1
460 DISP "ENTER YAXIS TIC MARKS(
    DB)"
470 INPUT M2
480 XAXIS -K,M1
490 YAXIS A,M2
500 REM LABEL XAXIS
510 LDIR 90
520 FOR X=A+2*M1 TO B STEP M1
530 MOVE X,-(K*.9)
540 LABEL VAL$(X)
550 NEXT X
560 REM LABEL YAXIS
570 LDIR 0
580 FOR Y=-K TO 0 STEP M2
590 MOVE A+(B-A)/20,Y
600 LABEL VAL$(Y)
610 NEXT Y
620 MOVE A+(B-A)/5,.2*(-K)
630 LABEL "BUTTERWORTH FILTER"
640 MOVE A+(B-A)/5,.25*(-K)
650 LABEL "FREQUENCY RESPONSE"
660 MOVE A+(B-A)/1.5,.3*(-K)
670 LABEL "N="&VAL$(N1)
```

```
680 MOVE A+(B-A)/2.2,.97*(-K)
690 LABEL "(MHZ)"
700 MOVE A+(B-A)/7,.5*(-K)
710 LABEL "(DB)"
720 MOVE A,-K
730 GOTO 210
740 END
```

Example 5-2:

Computer Inputs:

Select Filter	(R)eject
-3 dB Points (MHz)	
FMIN	300
FMAX	330
N ?	4
Sweep (MHz)	
Start	250
Stop	350
Inc.	2
Ultimate Attenuation (dB)	60
Graph ?	YES
Y axis max dB	70
X axis TIC (MHz)	10
Y axis TIC (dB)	10

The result of these computer inputs are shown in Fig. (5-15). The program will provide responses for all of the popular forms of this class of filter.

5.1.6 Modeling the Chebyshev Filter

This popular form of filter may be modeled in its theoretical form by the equation shown below:

$$A\,(\,dB\,) = 10\,\log_{10}\left[\,1 + \left[\,(\,\log^{-1}\,\frac{A_{max}}{10}\,) - 1\,\right]\cosh^2\left[\,n\,\cosh^{-1}\,(\,k\,)\,\right]\,\right]$$

$$(5\text{-}6)$$

where

A_{max} is the ripple in dB

n is the number of resonators for the bandpass and notch filter cases

n = the number of reactances in the low- and highpass cases

$k = b_3/b_x$ for the highpass and band reject cases

and

$k = b_x/b_3$ for the low- and bandpass cases

(b_3 is the 3 dB bandwidth and b_x is the bandwidth at frequency x. While this definition is not exact it is useful for k values of two or more.)

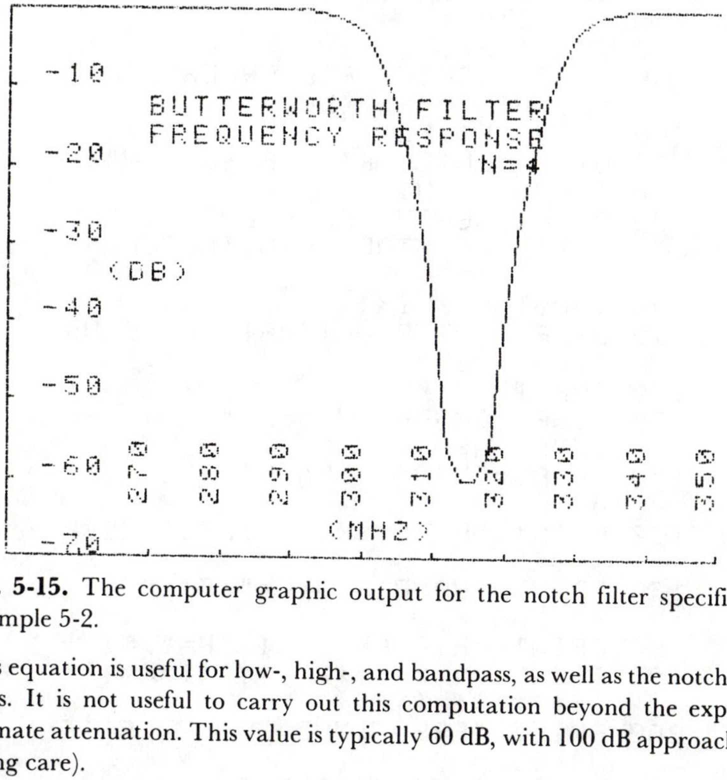

Fig. 5-15. The computer graphic output for the notch filter specified in Example 5-2.

This equation is useful for low-, high-, and bandpass, as well as the notch filter cases. It is not useful to carry out this computation beyond the expected ultimate attenuation. This value is typically 60 dB, with 100 dB approachable (using care).

The sharpness of the response is a function of the value of n and ripple. As these values increase, the shape factor is reduced and approaches unity. The responses predicted are approachable in practice; although some allowance must be made for departure in the practical sense.

This equation has been programmed in BASIC and is shown in Table 5-6. The program is interactive and prompting, with print or plot options. The user needs only to answer the questions asked. Some modifications may be necessary for machine variances.

Table 5-6.

Basic Program for the Print or Plot of the Chebyshev Filter Response for Low-,
High-, Bandpass and Notch Filter

```
10 SHORT F
20 DISP "CHEBYSHEV FILTER RESPO
   NSE"
30 DISP "SELECT FILTER TYPE (L)
   OW,(B)AND,(H)IGH, PASS OR (R
   )EJECT"
40 INPUT B$
50 DISP "DEFINE FILTER BAND WID
   TH;FMIN,FMAX,(MHZ)"
60 INPUT F1,F2
70 DISP "ENTER RIPPLE (DB),N"
80 INPUT R1,N1
90 DISP "DEFINE FREQUENCY SWEEP
   ; START, STOP, AND INCREMENT
    (MHZ)"
100 INPUT A,B,I
110 DISP "ENTER ULTIMATE ATTENUA
    TION (DB)"
120 INPUT U
130 DISP "GRAPH? YES NO"
140 INPUT A$
150 IF A$="YES" THEN 510 ELSE 16
    0
160 PRINT "CHEBYSHEV FILTER RESP
    ONSE"
170 PRINT "BAND WIDTH=";F1;"TO";
    F2;"MHZ"
180 PRINT "RIPPLE=";R1;"N=";N1
190 PRINT "*******************************
    ************"
200 PRINT "FREQ (MHZ)       ATTE
    N (DB)"
210 FOR F=A TO B STEP I
220 IF B$="B" THEN GOTO 260
230 IF B$="L" THEN GOTO 300
240 IF B$="H" THEN GOTO 370
250 IF B$="R" THEN GOTO 330
260 B1=2*ABS(F-((F2-F1)/2+F1)) @
      B3=F2-F1 @ X2=B1/B3
270 IF F<F1 THEN 290 ELSE 280
280 IF F>F2 THEN 360 ELSE 460
290 GOTO 400
300 B1=F @ B3=F2 @ X2=B1/B3
```

```
310 IF F<F2 THEN 460
320 GOTO 400
330 B1=2*ABS(F-((F2-F1)/2+F1)) @
    B3=F2-F1 @ X2=B3/B1
340 IF F>F2 THEN 460
350 IF F<F1 THEN 460
360 GOTO 400
370 B1=F @ B3=F1 @ X2=B3/B1
380 IF F>F1 THEN 460
390 GOTO 400
400 Y=N1*LOG(X2+(X2^2-1)^.5)
410 C1=(EXP(Y)+EXP(-Y))/2
420 R=10^(R1/10)-1
430 X3=10*LGT(1+R*C1^2*X2)
440 X3=ABS(100*X3+.5)/100
450 GOTO 470
460 X3=0
470 IF X3>U THEN X3=U
480 IF A$="NO" THEN PRINT TAB(1)
    ;F;TAB(19);X3 ELSE DRAW F,-X
    3
490 NEXT F
500 GOTO 890
510 GRAPH
520 GCLEAR
530 DISP "ENTER YAXIS MAX ATTN(D
    B)"
540 INPUT K
550 SCALE A,B,-K,0
560 DISP "ENTER XAXIS TIC MARKS
    (MHZ)"
570 INPUT M1
580 DISP "ENTER YAXIS TIC MARKS(
    DB)"
590 INPUT M2
600 XAXIS -K,M1
610 YAXIS A,M2
620 REM LABEL XAXIS
630 LDIR 90
640 FOR X=A+2*M1 TO B STEP M1
650 MOVE X,-(K*.9)
660 LABEL VAL$(X)
670 NEXT X
680 REM LABEL YAXIS
690 LDIR 0
700 FOR Y=-K TO 0 STEP M2
```

```
710 MOVE A+(B-A)/20,Y
720 LABEL VAL$(Y)
730 NEXT Y
740 MOVE A+(B-A)/5,.2*(-K)
750 LABEL "CHEBYCHEV FILTER"
760 MOVE A+(B-A)/5,.25*(-K)
770 LABEL "FREQUENCY RESPONSE"
780 MOVE A+(B-A)/5,.3*(-K)
790 LABEL "RIPPLE="&VAL$(R1)
800 MOVE A+(B-A)/1.5,.3*(-K)
810 LABEL "N="&VAL$(N1)
820 MOVE A+(B-A)/2.2,.97*(-K)
830 LABEL "(MHZ)"
840 MOVE A+(B-A)/7,.5*(-K)
850 LABEL "(DB)"
860 MOVE A,-K
870 GOTO 210
880 DRAW F,K3
890 END
```

Example 5-3:

Computer inputs for the Chebyshev filter program:

Filter	(H)ighpass
Bandwidth (MHz)	
F min	400
F max	10,000
Ripple (dB)	0.1
N	5
Sweep (MHz)	
Start	0
Stop	600
Increment	6
Ultimate Attenuation (dB)	65
Graph ?	YES
Y axis max (dB)	80
X axis TIC (MHz)	100
Y axis TIC (dB)	10

The computer output for this filter is shown in Fig. (5-16).

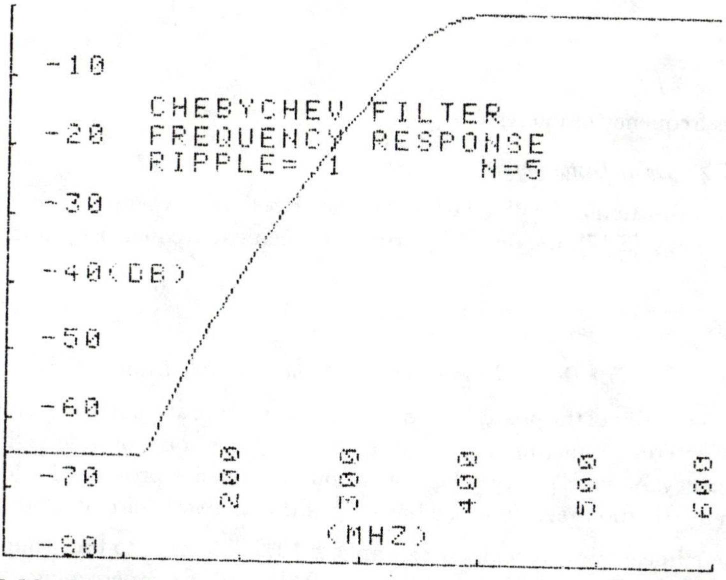

CHEBYCHEV FILTER
FREQUENCY RESPONSE
RIPPLE=.1   N=5

Fig. 5-16. Computer print-out for Example 5-3. This is a highpass Chebyshev filter with a ripple of 0.1 dB and $N=5$. The program is usable for the other popular filter types.

5.1.7 Distortion

Since nothing is perfect, the output of a device, circuit, or network differs from the input stimulus. This difference or distortion has several forms and definitions, which include:

Phase delay
Delay distortion
Envelope distortion, group delay, or envelope delay

Each of these will be described in the following paragraphs.

5.1.7.1 Phase Delay

Phase delay results from the time it takes a sinusoidal stimulus to pass through a circuit network or device. Because of this time delay (which is almost always frequency dependent) the output has a phase shift, which is a function of frequency. This characteristic is shown in a typical example, in Fig. (5-17). Here the phase delay may be computed from:

$$t_o = \frac{phase\ shift\ in\ radians}{\omega\ (radians/second)} = \frac{\theta}{\omega} \quad seconds \qquad (5-7)$$

where

$$\omega = 2\pi f$$

f is frequency of interest

5.1.7.2   Delay Distortion

Delay distortion is the phase delay difference between two frequencies. Referring to Fig. (5-17), the delay distortion between two frequencies f_1 and f_2 is:

$$t_d = \frac{\theta_2}{\omega_2} - \frac{\theta_1}{\omega_1} \quad (\textit{seconds}) \tag{5-8}$$

5.1.7.3   Envelope Delay, Group Delay, or Absolute Envelope Delay

The derivative of the phase shift *versus* frequency curve at a particular frequency of interest results in the rate of change of phase or group delay at that frequency. A plot of group delay, for various frequencies, provides the designer with information regarding the linearity of that network, circuit, or device.

Where linearity is critical, for example $\pm\, 10\%$, reference to the group delay curve will define the frequency limits available (within this percentage). Given group delay, the phase shift at any frequency may be computed from:

Phase shift = group delay (sec) $\cdot\, \omega$

$$\theta = \frac{d\theta}{d\omega} \cdot \omega \tag{5-9}$$

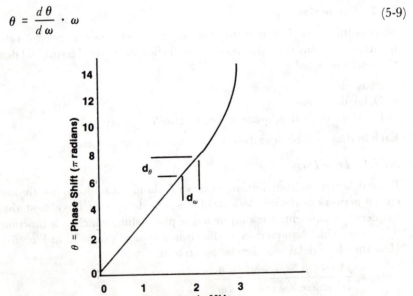

Fig. 5-17. An illustration of the various forms of definitions of distortion and delay.

Example: The group delay at 2 MHz is given as 1.825 μseconds; find the phase shift at this frequency.

1.825 $\cdot$ 10^{-6} $\cdot$ 2 π $\cdot$ 2 x 10^{-6} radians/seconds =
7.3 radians = 7.3 $\cdot$ 57.3°/radian = 418.3°

This example was taken from Fig. (5-17). (Note the correlation in the result.)

5.1.7.4   Relative Envelope Delay or Group Delay Relative

The group delay between two frequencies, where one is the reference, provides a measure of relative group delay in seconds.

Example 5-4 (Ref. Fig. (5-17)):

Phase delay at 1 MHz.

$$t_\theta = \frac{3.6\ \pi}{2\ \pi\ \cdot\ 1\ \cdot\ 10^6}\quad \frac{Rad}{Rad/sec} = 1.8\ \mu sec$$

Delay distortion from 1 to 2.5 MHz.

$$td = t_\theta\ (\ 1\ MHz\ ) - t_\theta\ (\ 2.5\ MHz\ )$$

$$= 1.8\ \mu sec - \frac{9.5\ \pi\ Rad}{2\ \pi\ \cdot\ 2.5\ \cdot\ 10^6\ Rad/sec.} = -0.1\ \mu sec.$$

Group delay at 2 MHz.

$$\tau_g = \frac{d\theta}{d\omega} = \frac{(\ 8 - 6.54\ )\ \pi\ Rad}{(\ 2.2 - 1.8\ )\ 2\ \pi\ 10^6\ Rad/sec} = 1.825\ \mu sec.$$

5.1.8   Computer Prediction of Group Delay for Butterworth and Chebyshev Filters

The designer may wish to have a print-out or plot of group delay for particular frequencies and bandwidths, without the laborious process of translation from the normalized curves contained in filter handbooks. Two sets of equations are presented for two popular filter types, the Butterworth and the Chebyshev. The equations are applicable for lowband and highpass responses and may be used for the readers' own predictions. For those who have access to machines programmable in BASIC, two programs are included. It may be necessary for the reader to modify these programs to cope with machine variances.

5.1.8.1   Butterworth Filter Group Delay

Group delay for the Butterworth class of filters may be approximated from:

$$\tau_g = \frac{1}{2\ \pi\ (\ F_{max} - F_{min}\ )}\ \left[\ 1 + \jmath\ \right]\ \sum_{k=1}^{n}\ \frac{|\ \sigma k\ |}{\sigma_k^2 + (\ x - w_k\ )^2} \qquad (5\text{-}10)$$

where

τ_g is the group delay in seconds

F_{max} is the upper –3 dB frequency (Hz)

F_{min} is the lower –3 dB frequency (Hz)

F is the frequency of computation (Hz)

n is the number of resonators for bandpass cases or the number of reactances for the lowpass case

J is 0 for lowpass and 1 for bandpass filters

$\sigma_k = \cos \theta_k$

$w_k = \sin \theta_k$

$$\theta_k = \frac{2k + n - 1}{n} \, 90°$$

$$x = \left| \frac{1}{F_{max} - F_{min}} \left[F - \frac{F_{max} \, F_{min}}{F} \right] \right|$$

For lowpass, enter a non-zero value for F_{min} such as 0.000001.

For bandpass filters enter the –3 dB points. The above equations are contained in the BUTTERWORTH GROUP DELAY PREDICTION program of Table 5-7. The program language is BASIC. Prompts are included in interactive form, with print or plot options.

Example: Two examples using this program are included for reader reference. The first example is a 100 MHz lowpass filter with five sections. The following parameters were entered:

L for lowpass filter

$N = 5$

$F_{max} = 100$

$INC = 2$ for the increment

YES for PLOT ?

10 ns for Y axis

10 for X axis TIC marks

1 for Y axis TIC marks

Table 5-7.

Program for the Prediction of Group Delay for Low- and Bandpass
Butterworth Filters

```
10  PRINTER IS 1
20  SHORT F,F1,F2,F3,T1,X,Y
30  DISP "BUTTERWORTH FILTER GRO
    UP DELAY"
40  DISP "CHOOSE LOW (L) OR BAND
    (B) PASS FILTER"
50  INPUT F$
60  IF F$="L" THEN J=0 ELSE J=1
70  DISP "ENTER NUMBER OF SECTIO
    NS"
80  INPUT N1
90  DISP "ENTER FMIN,FMAX,F INCR
    EMENT (MHZ)"
100 INPUT F1,F2,F3
110 DISP "GRAPH? YES NO"
120 INPUT B$
130 IF B$="YES" THEN 310 ELSE 14
    0
140 PRINT "BUTTERWORTH GROUP DEL
    AY"
150 PRINT "N=";N1
160 PRINT TAB(1);"MHZ";TAB(18);"
    NS"
170 FOR F=F1 TO F2 STEP F3
180 DEG
190 T1=0
200 FOR N=1 TO N1 STEP 1
210 A=(2*N+N1-1)/N1*90
220 O=ABS(1/(F2-F1)*(F-F2*F1/F))
230 S2=SIN(A)
240 C1=COS(A)
250 T=1000/(2*PI*(F2-F1))*(1+J)*
    ABS(C1)/(C1^2+(O-S2)^2)
260 T1=T1+T
270 NEXT N
280 IF B$="NO" THEN PRINT TAB(1)
    ;F;TAB(15);T1 ELSE DRAW F,T1
290 NEXT F
300 GOTO 670
310 GRAPH
320 GCLEAR
```

```
330 DISP "ENTER YAXIS MAX VALUE
    NANOSECONDS"
340 INPUT K
350 SCALE F1,F2,0,K
360 DISP "ENTER XAXIS TIC MARKS(
    MHZ)"
370 INPUT M1
380 DISP "ENTER YAXIS TIC MARKS(
    NANOSECONDS)"
390 INPUT M2
400 XAXIS 0,M1
410 YAXIS F1,M2
420 REM LABEL XAXIS
430 LDIR 90
440 FOR X=F1+M1 TO F2 STEP M1
450 MOVE X,K/20
460 LABEL VAL$(X)
470 NEXT X
480 REM LABEL YAXIS
490 LDIR 0
500 FOR Y=M2 TO K STEP M2
510 MOVE F1+(F2-F1)/20,Y
520 LABEL VAL$(Y)
530 NEXT Y
540 MOVE F1+(F2-F1)/5,.9*K
550 LABEL "BUTTERWORTH FILTER"
560 MOVE F1+(F2-F1)/5,.85*K
570 LABEL "GROUP DELAY"
580 MOVE F1+(F2-F1)/5,.8*K
590 LABEL "N="&VAL$(N1)
600 MOVE F1+(F2-F1)/2.2,.2*K
610 LABEL "(MHZ)"
620 MOVE F1+(F2-F1)/7,.5*K
630 LABEL "(NS)"
640 MOVE F1,0
650 GOTO 170
660 DRAW F,T1
670 END
```

The result of this program is shown in Fig. (5-18). A second example is shown for a bandpass case. This example retains the same filter bandwidth of 100 MHz; to show the doubling of the delay for a bandpass filter (for a given bandwidth as compared to the lowpass case). The entries are as follows:

B for bandpass filter

$N = 5$

$F_{min} = 300$

F_{max} = 400

INC = 2 for the increment in frequency

YES for PLOT ?

20 *ns* for the Y axis

10 for X axis TIC marks

2 for the Y axis TIC marks

The results are shown in Fig. (5-18). Note the doubling of the lowpass values for the bandpass filter with the same bandwidth.

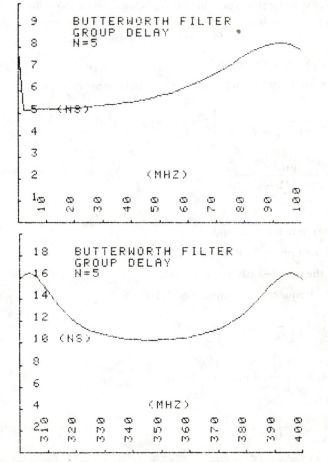

Fig. 5-18. Examples of lowpass and highpass filter group delay predictions using the computer program of Table 5-7. Note the filter bandwidths are the same, and the bandpass filter group delay is twice that of the lowpass case. Also note the program inaccuracies near zero frequency for the lowpass case.

Please note that the program for the lowpass case is inaccurate near zero frequency. The true characteristics would be an extrapolation of the curve from the value (from right to left). In other words, the characteristic is flat near the zero frequency point.

It is suggested that a print be selected (NO to PLOT?) before a PLOT is attempted, to secure the scale factors. The print output is not constrained and the actual values will be output. Once this is secured, a plot can be executed with the appropriate scale factors.

The program statement 10 PRINTER IS 1, should be deleted for the print option. This statement was only included to save paper in the print-out. Once a satisfactory solution is achieved, a print condition can be exercised. Alternatively, COPY could be used for a hard copy of the screen.

5.1.8.2   *Chebyshev Filter Group Delay*

Chebyshev filter group delay may be computed from:

$$\tau_g = \frac{1}{2\pi(F_{max} - F_{min})}\left[1 + \frac{F_{max}F_{min}}{F^2}\right] \cdot$$

$$\sum_{k=1}^{n} \frac{|\sigma_k|}{\sigma_k^2 + (\Omega - w_k)^2} \tag{5-11}$$

where

τ_g is group delay in seconds

F_{max} is the upper −3 dB frequency (Hz)

F_{min} is the lower −3 dB frequency (Hz)

F is the frequency of computation (Hz)

$$\sigma_k = \sinh\left[\frac{1}{n}\sinh^{-1}\frac{1}{a}\right]\sin\theta_k$$

$$w_k = \cosh\left[\frac{1}{n}\sinh^{-1}\frac{1}{a}\right]\cos\theta_k$$

$$a = \left(\log^{-1}\frac{A_{max}}{10} - 1\right)^{1/2}$$

$$\Omega = \left|\frac{1}{F_{max} - F_{min}}\left[F - \left(\frac{F_{max}F_{min}}{F}\right)\right]\right|$$

$$\theta_k = \frac{2k - 1}{n} \ 90°$$

$$\sinh^{-1} x = \ln [x + (x^2 + 1)^{1/2}]$$

where

n is the number of resonators (bandpass and notch) and the number of reactances (low- and highpass)

$$\sinh x = \frac{e^x - e^{-x}}{2}$$

$$\cosh x = \frac{e^x + e^{-x}}{2}$$

This program approximates the actual value at mid-band and the band edges but tends to be low in other areas. It is still useful in approximate terms.

A_{max} is ripple (dB)

For lowpass filters $F_{min} = 0$

For highpass filters $F \gg F_{max}$

The above equations which are quite similar to those of the Butterworth class, are contained in the Chebyshev Group Delay Prediction Program of Table (5-8), and include prompts. The program is interactive and allows the selection of print or plot options.

The references for this section are [1] and [2].

Example 5-5:

Ripple(dB)	0.5
N (sections)	4
–3 dB Points (MHz)	
Min	850
Max	950
Increment	2
GRAPH ?	YES
Y axis max value *ns*	24
X axis TIC MHz	10
Y axis TIC *ns*	2

The results of these inputs to the computer, using the Chebyshev program for group delay, results in the output of Fig. (5-19).

Table 5-8.

Program for the Prediction of Chebyshev Filter Group Delay

```
10  SHORT F,F1,F2,F3,T1,X,Y.
20  DISP "CHEBYSHEV FILTER GROUP
    DELAY"
30  DISP "ENTER RIPPLE(DB),NUMBE
    R OF SECTIONS"
40  INPUT R,N1
50  DISP "ENTER FMIN,FMAX,F INCR
    EMENT (MHZ)"
60  INPUT F1,F2,F3
70  DISP "GRAPH? YES NO"
80  INPUT B$
90  IF B$="YES" THEN 290 ELSE 10
    0
100 PRINT "CHEBYSHEV GROUP DELAY
    "
110 PRINT "RIPPLE=";R,"N=";N1
120 PRINT TAB(1);"MHZ";TAB(18);"
    NS"
130 FOR F=F1 TO F2 STEP F3
140 DEG
150 T1=0
160 FOR N=1 TO N1 STEP 1
170 E=(10^(R/10)-1)^.5
180 S1=1/N1*LOG(1/E+((1/E)^2+1)^
    .5)
190 A=(2*N-1)/N1*90
200 O=ABS(1/(F2-F1)*(F-F2*F1/F))
210 S2=(EXP(S1)-EXP(-S1))/2*SIN(
    A)
220 C1=(EXP(S1)+EXP(-S1))/2*COS(
    A)
230 T=1000/(2*PI*(F2-F1))*(1+F2*
    F1/F^2)*ABS(S2)/(S2^2+(O-C1)
    ^2)
240 T1=T1+T
250 NEXT N
260 IF B$="NO" THEN PRINT TAB(1)
    ,F;TAB(15);T1 ELSE DRAW F,T1
270 NEXT F
280 GOTO 670
```

```
290 GRAPH
300 GCLEAR
310 DISP "ENTER YAXIS MAX VALUE
    NANOSECONDS"
320 INPUT K
330 SCALE F1,F2,0,K
340 DISP "ENTER XAXIS TIC MARKS(
    MHZ)"
350 INPUT M1
370 DISP "ENTER YAXIS TIC MARKS(
    NANOSECONDS)"
380 INPUT M2
385 XAXIS 0,M1
390 YAXIS F1,M2
400 REM LABEL XAXIS
410 LDIR 90
420 FOR X=F1+M1 TO F2 STEP M1
430 MOVE X,K/20
440 LABEL VAL$(X)
450 NEXT X
460 REM LABEL YAXIS
470 LDIR 0
480 FOR Y=M2 TO K STEP M2
490 MOVE F1+(F2-F1)/20,Y
500 LABEL VAL$(Y)
510 NEXT Y
520 MOVE F1+(F2-F1)/5,.9*K
530 LABEL "CHEBYCHEV FILTER"
540 MOVE F1+(F2-F1)/5,.85*K
550 LABEL "GROUP DELAY"
560 MOVE F1+(F2-F1)/5,.8*K
570 LABEL "RIPPLE="&VAL$(R)
580 MOVE F1+(F2-F1)/1.5,.8*K
590 LABEL "N="&VAL$(N1)
600 MOVE F1+(F2-F1)/2.2,.2*K
610 LABEL "(MHZ)"
620 MOVE F1+(F2-F1)/7,.5*K
630 LABEL "(NS)"
640 MOVE F1,0
650 GOTO 130
660 DRAW F,T1
670 END
```

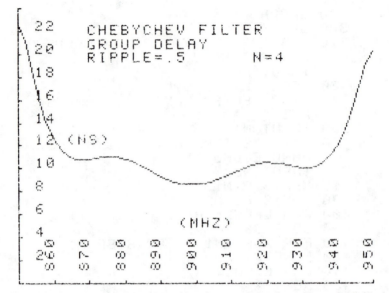

Fig. 5-19. The computer prediction of group delay for a filter of the Chebyshev class with a ripple of 0.5 dB and four resonators. See Example 5-5.

5.2 MIXERS

The mixer is the key to superheterodyne design. It may be considered a three port component. There are two input ports. These consist of the received signal and local oscillator ports. The output is the intermediate frequency port. The mixer (sometimes referred to as a multiplier, though it is not in the strict sense) outputs two principal signals, which consist of the sum and the difference of the two inputs. In practice, only one of these is selected as the desired signal. This selection is done by appropriate filtering of the mixer's output. Either output is acceptable as the signal, depending on the mixer mode (sum or difference).

Mixers may take on several different forms, which may employ: diodes, transistors of bipolar or field effect types, *et cetera*. The three basic mixer forms include:

single ended
balanced
double balanced

These configurations are shown in Fig. (5-20). Variations include the use of transistors, several diodes in series, either of these in series with resistors and capacitor balancing, or any combination of these. Higher performance balanced mixers usually employ baluns at the RF and LO ports when the inputs are unbalanced.

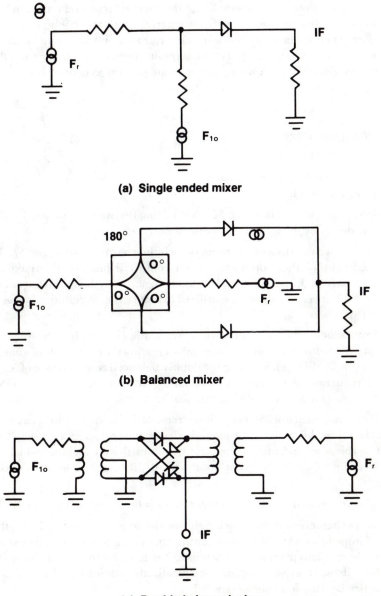

(a) Single ended mixer

(b) Balanced mixer

(c) Double balanced mixer

Fig. 5-20. The three basic mixer forms: (a) single ended, (b) single balanced, (c) double balanced.

The single ended mixer is the simplest of the various forms and is used in low performance radios, such as in the home entertainment and hobby field. It is the poorest performer from the spurious product point of view. Spurious products are those mixer outputs which result in something other than the desired converted signal. A mixer will generate outputs as defined by:

$$|M F_{10} \pm N F_r|$$

where

   M and N are integers including 0

and

   F_{10} is the local oscillator frequency
   F_r is the received frequency

The desired output occurs where $M = N = 1$ using the appropriate sign for the mixer mode (+ for the sum and − for the difference).

There can exist a multiplicity of products, including harmonics of F_r and F_{10}. To avoid as many of these products (which could fall into the intermediate frequency passband, and appear as legitimate signals even though all but one of the $M = N = 1$ are not), balancing is utilized to reduce the magnitude of some of these terms.

The simplest of the balanced mixers is shown in Fig.(5-20b). Here the spurious population is reduced through balance by decreasing the magnitude of some of the terms by 20 dB or so. By using the doubly balanced configuration of (c), a further reduction of the spurious population is secured. This is the most popular class of mixer which is used in quality products.

A further consequence of balance is the increase of the power handling capability of the mixer, because two or more diodes or active devices are used. The result of this is an increase of the local oscillator drive capability which, if exploited, increases the intercept point and −1 dB compression point of the mixer.

The relative merits of the three configurations is tabulated in Table (5-9).

Mixers are characterized by using purely resistive terminations matched to the mixer's impedance. Any other non-resistive matching results in an increased level of the spurious population. Therefore, it is desirable to provide the mixer with broadband resistive matching. This idealized condition is seldom realized in practice because it is usually ignored.

The mixer, local oscillator, and input signal ports are easily matched using broadband amplifier drivers. Filters' driving mixers provide a match only within their bandpass and should be avoided when high performance is desired.

Table 5-9.

Relative Mixer Performance

	Single Ended	Single Balanced	Double Balanced
		Class	
Spurious Density	4X	2X	1
L.O. Drive Typ. dBm	7	7 to 10	10 to 27 or more
Port to Port Isolation dB	—	10 to 20	20 to 50
–1 dB Compression dBm	0	0 to 4	4 to 23
Intercept Point dBm	10	10 to 15	20 to 33

The same holds true for the IF output port. The IF output port should see a broadband match such as is provided by a broadband amplifier's input impedance. An alternative solution is to use two filters with parallel inputs. The IF bandpass filter's output is fed to the IF amplifier. This filter provides a match to the mixer within its bandpass. Elsewhere the impedance is not matched. By adding a notch filter in parallel at the input and terminating this filter resistively, the input impedance of this pair of filters will be essentially flat over a broadband. This filter can provide a match to the mixer IF port everywhere except within the notch, where it has a higher impedance. By making the notch bandwidth equal to the desired IF bandwidth, a broadband match is achieved. These mixer matching methods are shown in Fig. (5-21).

5.2.1 Specialized Mixers and Applications

5.2.1.1 Termination Insensitive Mixers

A termination insensitive mixer results when the configuration of Fig. (5-22) is employed.

Mixers of this type are popular in high performance designs. They are capable of handling strong received signals with good spurious performance without exercising care in impedance matching.

5.2.1.2 Image Rejection Mixers

In certain wideband applications it may not be practical to utilize preselection filtering of the RF input signals. In such cases, the image response could appear as a legitimate signal. A degree of image rejection (20 to 30 dB) is achievable by using the configuration of Fig. (5-23).

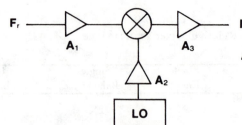

F$_r$ ──▷── ⊗ ──▷── IF Filter

A$_1$ A$_3$

A$_2$

LO

**A$_1$, A$_2$, A$_3$ are broad-
band amplifiers**

(a) Broadband amplifier matching of a mixer.

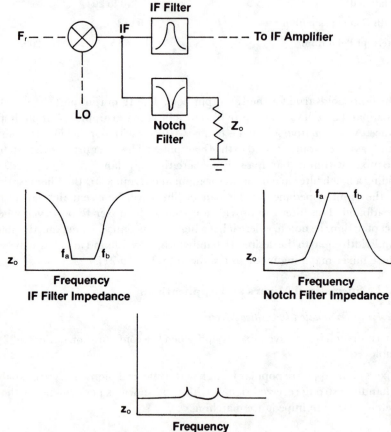

(b) IF Port Matching Using a Notch Filter.

Fig. 5-21. Methods of broadband matching of a mixer IF Port.

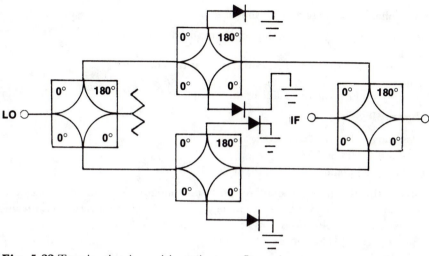

Fig. 5-22 Termination insensitive mixer configuration.

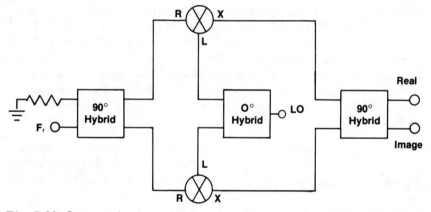

Fig. 5-23. Image rejection mixer configuration capable of 20 to 30 dB of rejection.

The degree of rejection depends upon circuit balance. Not only must the amplitudes of the converted signals at the final combiner be equal, but the phase must also be matched. The severity of this problem is addressed in reference [3]. Amplitude balance is given as:

$$(I + Q) / (I - Q)$$

Example 5-6:

If $I = 1.1\ Q$

then the balance would be 21 or 26.44 dB and the two signals would have to be amplitude matched to .83 dB.

The phase balance severity is shown to be:

$\cos \theta = 1 - c^2/2a^2$ assuming ideal amplitude balance

Where
$\quad a$ and b are the two converted signals at the combiner
and
$\quad a = b$ and c is the third side of the triangle.

Example 5-7:

Assume $a = 1$ and $c = .1$ then
$\quad \cos \theta = .995$ or $5.73°$
$\quad$ and the balance is $2a/c$ or 20 or 26 dB.

One of the difficulties encountered with this configuration is maintaining gain and phase balance over broad frequency bands. However, for certain relatively narrow band applications, this represents an ideal solution.
See reference [3] for more information.

5.2.1.3 Harmonic Mixers

Often at microwave frequencies a mixer is operated at product orders other than 2. That is, $M + N \neq 2$. More specifically, $M F_{lo} \pm F_r = IF$ where $N = 1$ and the sign as applies. Thus the mixer utilizes harmonics of the local oscillator to combine with the fundamental of the received frequency to generate the intermediate frequency. While effective, the conversion loss is necessarily high, directly increasing with M for single ended mixers. Balanced mixers are seldom used especially where the LO port is balanced, unless only odd harmonics of F_{lo} are used.

Typically LO harmonic numbers of 2 through 10 are common and in some special cases up to 60. In general, a good mixer is a poor harmonic mixer. Therefore the balanced mixer is a poor choice because harmonic responses are suppressed.

Biasing also has the effect of reducing the required LO drive which may be reduced to 0 or −10 dBm in starved operation.

The conversion loss of the harmonic mixer is on the order of:

HARMONIC OF LO	CONVERSION LOSS, −dB
2	12
3	13
4	15
5	17
6	22
7	15
8	25

The noise figure is the conversion loss. Consequently, harmonic mixers are not used in a high sensitivity design.

5.3 LINEAR AND NON-LINEAR AMPLIFIERS

A linear amplifier faithfully reproduces all amplitude variations of the input signal at its output. A non-linear amplifier reproduces the input signal amplitude variations in a non-linear manner. The non-linearity may be a threshold which must be exceeded by the input signal before an output occurs. It may take the form of a linear amplifier operated in saturation, or an amplifier with a non-linear input/output relationship (such as a log amplifier).

There is definite need for each category of amplification. The linear amplifier can, of course, process any and all signals; however, it may not be economical to do so.

The application rules are as follows: where the information to be processed is contained in amplitude variations of the signal, the signal must be amplified linearly. (This does not include on/off keyed signals.) Failure to follow this rule will result in severe distortion or complete loss of information. Examples of some signals requiring linear amplification are shown in Table 5-10.

Where information to be processed is contained in phase, frequency, a combination of the two, pulsewidth, or position, non-linear amplification may be used.

Table 5-10.
Popular Signal Forms Requiring Linear Amplification

Double sideband amplitude modulation (DSB AM or AM)

Double sideband suppressed carrier amplitude modulation
  (DSB-SC AM)

Single sideband amplitude modulation
  (SSB-AM)

Multilevel pulse amplitude modulation
  (M PAM-AM)

Amplitude shift keying
  (ASK)

In certain applications, such as limiting amplification, the use of the non-linear form is mandatory. Some of the more popular forms of signals which may be amplified by non-linear means are shown in Table 5-11.

Table 5-11.

Popular Signal Forms Which may be Processed by Non-Linear Amplification

Continuous wave interrupted carrier (CW)

On-off keying (OOK)

Pulsewidth or duration modualtion (PWM or PDM)

Phase modulation (PM)

Multiphase modulation (digital includes biphase, four or eight phase, *etc.*)

Frequency modulation (FM)

Frequency shift keying (FSK)

5.3.1 Limiter

A limiter is a signal processor which has upper and lower bounds that limit the excursions of a function of time in the amplitude axis. The idealized transfer function of a limiter is shown in Fig. (5-24).

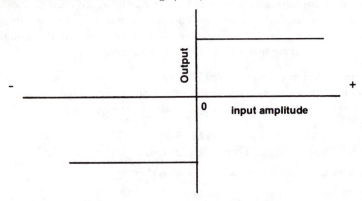

Fig. 5-24. Idealized limiter transfer characteristic.

This transfer function is known as a hard limiter because the action is definite and invariant. Once the limiting threshold is reached, any increase of input amplitude does not allow any further increase in output above that threshold. Departure from this characteristic results in what is known as soft limiting. The distinction has no well-defined boundary.

The behavior of a limiter is illustrated in Fig. (5-25).

Limiters are used to remove amplitude perturbations from a signal and in no way do they disturb the crossings of a signal in the time axis. Limiters should not be used to process signals where the message is modulated onto the carrier in the amplitude axis, such as AM, because the modulation would be removed. Where the message is modulated onto a carrier in quadrature to the amplitude axis, such as FM, limiting will not disturb the message. The limiting process causes the generation of harmonics of the signal frequency because the sinusoidal carrier is converted to a form of the square wave. Limiting also is effective in the removal of AM noise from a signal, as is obvious.

Many forms of FM demodulators require limiting of their input signals to maintain constant output with signal strength. Examples of this are the discriminator and the slope detector.

Limiters are simply devices which go into saturation on overload. One simple configuration is two diodes in parallel with one being reversed in series, with a current limiting resistor. The voltage across the diode would be limited to ± 0.6 volts for silicon diodes. Other forms are amplifiers with low supply voltage. The amplifier output cannot exceed the supply voltage. Most limiters are of this form.

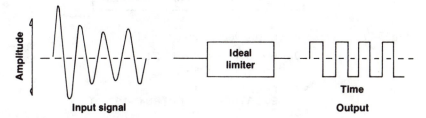
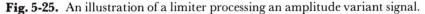

Fig. 5-25. An illustration of a limiter processing an amplitude variant signal.

5.3.2 Successive Detection Log Amplifiers

A logarithmic amplification characteristic may be desirable in applications where signal level information is required, or where instantaneous outputs are required from a wide dymanic range of signals. Linear amplifiers designed for threshold signals with gains of 100 dB (such as an IF amplifier) may have an output capability of several volts at threshold and a saturation level several times that. Therefore, only a slight increase in input level will cause saturation of the amplifier. Most such amplifiers have a headroom of only 3 to 6 dB. AGC is one solution for saturation but because it is generally non-linear and has a finite time constant, it cannot be used with any accuracy where instantaneous outputs are required or where signal strength measurements are required.

The use of a cascade of amplifiers, where the output of each stage is detected and all detector outputs are summed, results in a quasi-logarithmic, characteristic of the form:

$$e_o = k \log e_{in}$$

where k is a scale factor.

The configuration is shown in Figures (5-26) and (5-27), together with a computer-simulated output which assumes ideal linear detection. The result clearly shows the logarithmic characteristic with the deviation from the ideal. Because of the non-linear input / output characteristic, linear amplitude modulated signals will suffer from amplitude distortion. The main application of logarithmic amplifiers are in: pulsed signal systems, spectral analysis, filter characteristic measurements, *et cetera*.

Fig. 5-26. Successive detection logarithmic amplifier.

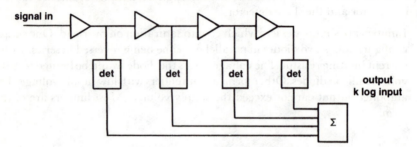

```
LOG AMPLIFIER ANALYSIS
NUMBER OF STAGES IS 5

STAGE GAIN IS 10
STAGE SATURATED OUTPUT IS 10
VOLTS

TOTAL GAIN IS 100 DB

INPUT SIGNAL IS .001 TO 1000
MILIVOLTS
```

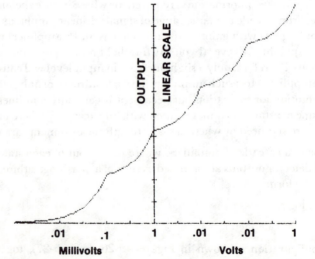

Fig. 5-27 A computer-simulated output for a successive detection log amplifier.

REFERENCES

[1] Schreiber, Heinz H., "Phase and Time Delay of Butterworth and Chebyshev Filters," *Microwaves,* March 1965, page 14.

[2] Nicholson, B.F., "The Practical Design of Interdigital and Comb-line Filters," *The Radio and Electronic Engineer,* July 1967, page 39.

[3] Gorwara, Asok K., "Phase and Amplitude Balance: Key to Image Rejection," *Microwaves,* October 1972, page 64.

6

SPECIALIZED RECEIVER APPLICATIONS

Other than the reception of communication signals for information exchange, the second most important application is the surveillance of sections of the RF spectrum in search of signals with particular characteristics. This is done largely in electronic warfare applications. The major variables which are involved are one or more of the following:

Frequency
Direction of the emitting source or signal spatial position
Time or duty cycle
Modulation characteristics
Signal strength, *et cetera*

These receiving systems consist of three major parts:

Antenna(s) and distribution system(s)
The receiver(s)
The signal processor

The antenna system may be called upon to provide directional information to the emitter and could consist of a single movable array, or a multiplicity of fixed antennas whose signal strengths (and/or times of signal arrival) are compared for signal positional computation. The latter may use one or more receivers.

The receiver may be called upon to provide the frequency of the emitter for jamming or logging purposes. Additionally, information regarding the signal strength of the emitter (at the receiving site) plus the nature and content of the modulation may be required.

All of this information is fed to a processor which may store and/or compare these data, with a reference for decision making purposes or analysis.

There are two major categories of receivers used for this purpose; included are narrowband (less than one spectral octave) and the wideband (covering one or more octaves). Included in the narrowband group are:

Scanning superheterodyne
Linear

Smart scan
Microscan or compressive receiver

The wideband group includes:

Channelized receivers
Crystal video
Tuned radio frequency
Instantaneous frequency measurement (IFM) receivers

The principal advantage of the wideband receiver group is their non-scanning operation, which permits instantaneous handling of frequency agile signals. A disadvantage is their spectral resolution. Each of the above will be discussed in the following sections.

6.1   DEFINITIONS: SCANNING SUPERHETERODYNE

Linear

The spectrum analyzer is an example of a superheterodyne structured receiver, whose local oscillator and preselector are tuned, tracked, and linearly swept over a range of frequencies.

Smart Scan

This is a special case of the linear type where the desired signal population is known and where the sweep is restricted to spectral zones. The advantage of this technique is the lower scan time.

Compressive or Microscan Receiver

This is another variation of the swept superheterodyne with an extremely fast sweep time (typically one microsecond). To resolve a signal without amplitude degradation, a wide IF bandwidth is required. The output of the wideband filter is an FM chirp which is fed to a compressive delay line, whose time delay is frequency dependent and matched to the sweep. This compresses the signal, concentrating the signal power while the noise is not concentrated, resulting in an enhanced signal to noise ratio.

Instantaneous Frequency Measurement (IFM) Receiver

This is a receiver which utilizes a wideband discriminator usually one or more GHz wide to measure frequency. Since it is a non-scanning system, its operation is instantaneous.

Channelized Receiver

The desired spectrum to be monitored is broken into cells, usually contiguous, though not necessarily so. Each of these cells is provided with receivers whose detection bandwidths are equal to that of the cells. Such a system is non-scanning and is therefore an instantaneous wideband type. Frequency measurement resolution is necessarily restricted to one cell width.

6.2   SCANNING SUPERHETERODYNE RECEIVER

A conventional superheterodyne receiver (when provided with a swept local oscillator and a wideband or tracking preselector) becomes a scanning superheterodyne receiver.

The scan speed is limited by the IF bandwidth. Narrow IF bandwidths require a slower sweep speed, or a loss of sensitivity will result. This relationship is:

$$S_l = \left[\frac{\Delta F^2}{5.282\, t^2 B^4} + 1 \right]^{-0.25} \tag{6-1}$$

where

    ΔF is the sweep width
    t is the time of one sweep
    B is the 3 dB bandwidth of the receiver

Bandwidth governs the noise floor level or threshold of detection for positive signal to nosie ratios. It also governs the resolution of the system. An appropriate selection of parameters for specific performance must be made. A typical scanning superheterodyne receiving system is shown in Fig. (6-1).

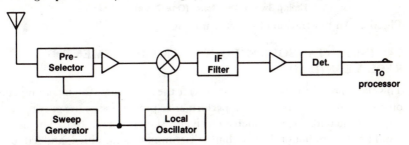

Fig. 6-1. A basic scanning superheterodyne receiver.

6.3   SMART SCAN RECEIVER

The smart scan receiver is a variation of the scanning superheterodyne receiver, which speeds up the system by omitting those portions of the spectrum which are of no interest. To illustrate, reference is made to Fig. (6-2). Shown is the local oscillator waveform, which is swept only through the frequency ranges of *F2* to *F3* and *F4* to *F5*. The remainder of the range (such as *F1* to *F2*, *F3* to *F4* and *F5* to *F6*) is skipped by increasing the sweep speed as much as possible. The result is that there will be no output of useful signals in these unwanted zones. The principal advantage of this technique is a lower acquisition time of the desired signals.

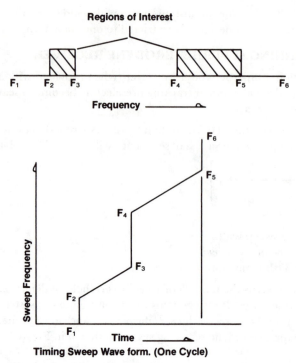

Fig. 6-2. An illustration of smart scanning.

6.4 INSTANTANEOUS FREQUENCY MEASUREMENT RECEIVERS (IFM)

This class of receivers is designed to provide the frequency of an incoming signal on an instantaneous basis. Such performance precludes the use of swept techniques. One realization of such a receiver is shown in Fig. (6-3). The signal is band limited to that of the wideband discriminator, amplified, and fed to two channels. These are the signal presence and the frequency measuring channels. The signal presence channel serves to detect the presence of a signal, and then keys a signal processor. The second channel consists of a delay line and a phase discriminator, driven from a power divider. The discriminator is sometimes called a correlator network. The arrangement is shown in Fig. (6-4).

The phase discriminator consists of two input ports which are a 180° and a 90° hybrid, which in turn drive four diode detectors through two 90° hybrids, as shown in Fig. (6-5). The outputs of the four detectors produce four outputs whose magnitude and phase are dependent upon the phase difference of the two inputs to the correlator. These are represented by:

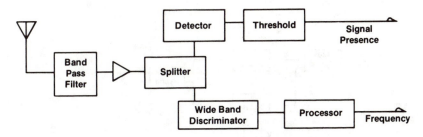

Fig. 6-3. An illustration of an instantaneous frequency measuring receiver.

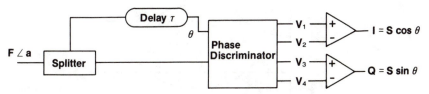

Fig. 6-4. The discriminator used in the IFM receiving system.

$V_1 = (A^2 + B^2) + 2AB \cos \theta$
$V_2 = (A^2 + B^2) - 2AB \cos \theta$
$V_3 = (A^2 + B^2) + 2AB \sin \theta$
$V_4 = (A^2 + B^2) - 2AB \sin \theta$

By differentially comparing V_1 and V_2, and also V_3 and V_4, the term $(A^2 + B^2)$ is eliminated, leaving two signals:

$I = S \cos \theta$
$Q = S \sin \theta$

These signals are furnished to a processor which vectorially sums the I and Q signals, producing a magnitude component proportional to signal power and an angle proportional to phase. The delay line shown in Fig. (6-4) establishes a phase θ, which is a function of frequency, since time and frequency are related. This relationship is linear, resulting in the capability of measuring frequency with good accuracy. To avoid ambiguity θ must be less than 360° over the frequency band of interest.

The delay line is theoretically errorless, however the correlator is not (because of its complexity and tracking errors). Typically, a ±6° error is realized. For a 1 to 2 GHz system this results in a ±17 MHz error possibility.

The correlator output is usually digitally processed. Should greater accuracy be required, four correlators may be used with 4^n weighting to provide 10 bit

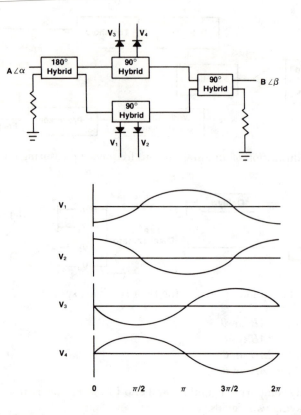

Fig. 6-5. Correlator configuration and output phase relationships.

accuracy, as shown in Fig. (6-6). The delay lines are t for D_1; $4t$ for D_2; $16t$ for D_3 and $64t$ for D_4. The respective resolutions are 1, $1/4$, $1/16$, and $1/64$. Thus, the system's resolution is 64 times better than that of one correlator system. The ambiguity is still determined by the delay line associated with D_1. A four-correlator system will have a resolution of one part in 1024. Therefore, for a 1 to 2 GHz band, the resolution is ~1 MHz. The dynamic range of the IFM is typically limited by the correlator to 30 dB.

6.5 MICROSCAN (COMPRESSIVE) RECEIVER

This variation of the swept superheterodyne (Fig. 6-7)) results in fast signal acquisition. To accomplish this, the sweep speed is dramatically increased to cover the entire tuning range in a microsecond or so. As pointed out in section 6.2, the signal amplitude loss would be very great unless a very wide IF bandwidth is used. To overcome this, the signal (which is a FM chirp) is fed to a compressive filter which has a time delay matched to the sweep. This delay is a function of frequency. For a positive or increasing frequency IF chirp, the time

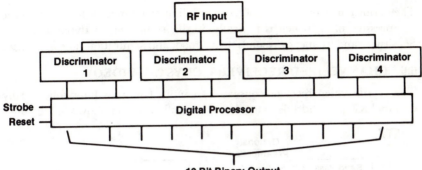

Fig. 6-6. Multiple discriminator arrangement for improved resolution.

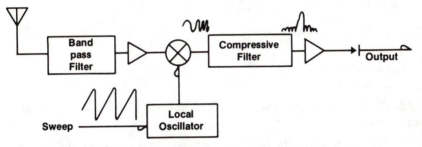

Fig. 6-7. Compressive (microscan) receiver block diagram.

delay of the compressive filter is negative with increasing frequency. This characteristic causes the low frequency end of the chirp (the first to arrive in this case) to be delayed most, while the higher frequencies are progressively delayed less. The result is a compression of the IF signal energy into a narrow pulse. The associated noise is not correlated and is therefore unaffected in density, whereas the signal energy is enhanced. The result is a compression gain of the signal, which is the time bandwidth product of the compressive filter.

Example 6-1:

Given

Tuning range   1 to 2 GHz
Time of one sweep   1 microsecond

Then

$B = |\,1 - 2\,\text{GHz}\,| = 10^9\,\text{Hz}$

The compression factor is:

$10^{-6} \times 10^9 = 10^3 = 30\,\text{dB}$

The output pulsewidth is:

1 microsecond $/\ 10^3 = 1$ nanosecond

The performance of this class of receiver is limited by the side lobe level of the compressive filter, which is typically 30 dB, restricting its dynamic range. Sensitivity and frequency resolution are good, and acquisition time is ~ 10 μs.

6.6 CHANNELIZED RECEIVERS (CRYSTAL VIDEO)

The crystal video receiver consits of an input bandpass filter followed by optional RF gain and a detector. A typical configuration is shown in Fig. (6-8).

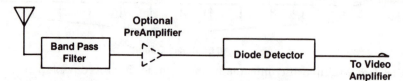

Fig. 6-8. Basic crystal video receiver.

Three of the most popular detectors are the:

Schottky diode
Tunnel diode
Point contact diode

Each diode has unique features which make it useful for a particular application.

The Schottky diode has the highest tangential signal sensitivity, output, and burnout rating.

The tunnel diode does not require bias. It has the best response time, lowest output video resistance, and the best thermal stability.

The point contact detector provides the best match, requires no bias, and has the best frequency response.

Of the three, the Schottky diode is the most popular, largely because of its higher sensitivity. Typically this diode is capable of a TSS of –50 to –52 dBm in a 2 MHz video bandwidth, and of a video amplifier noise figure of 3 dB. This can be improved through the addition of an RF amplifier.

The crystal video receiver is most often used at microwave frequencies because of its broadband nature. It is ideal for channelized applications where several receivers are used to cover adjacent frequency bands. The response is instantaneous for signal presence because of its non-scanning nature; it is also capable of a degree of frequency measurement resolvable to its bandwidth.

An example of channelized receiving is shown in Fig. (6-9). Here the 1 to 2 GHz frequency range is divided into 10 bands, each of which is covered by a crystal video receiver (preceded by an appropriate bandpass filter). The frequency resolution would be:

$(1 - 2\,\text{GHz})/10 = 100\,\text{MHz}$

Because the bandpass filters are not ideal, there is a zone of overlap at the band edges with adjacent filters. Should this be of concern, adjacent channel output amplitude comparison would resolve this problem.

The preamplifier is usually of bipolar, tunnel diode, or GaAs FET design. Where low power consumption is a major concern, tunnel diode and, more recently, GaAs FET amplifiers are used.

The function of the preamplifier is to provide sufficient gain to overcome detector noise and become as dominant as the noise source. Typically, 40 dB of gain is used for this purpose.

Because of the non-AGC'd design, the crystal video receiver has a low dynamic range. This is particularly important when preamplification precedes the detector.

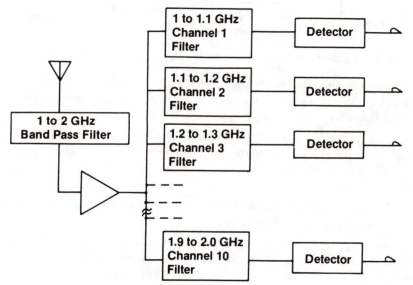

Fig. 6-9. Basic 10 channel channelized crystal video receiver.

6.7 BRAGG CELL RECEIVER

The Bragg cell receiver is an electrooptical spectrum analyzer. The basic configuration is shown in Fig. (6-10). The Bragg cell is a processor which is furnished with two inputs, one of which is from a wideband receiver through an RF acoustic coupler and the second, an optical beam from a laser source. The cell is acoustically and optically transparent with photon-phonon relationships directly related to its inputs. This results in a variation of the refractive index of the cell, which is linearly related to signal frequency.

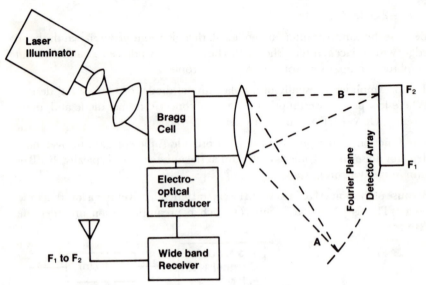

Fig. 6-10. An illustration of a basic Bragg cell receiver.

With no signal input the laser beam remains at *A*, which is off the detector array surface. An increasing RF input signal frequency causes the beam to deflect toward *B* in a linear relationship.

The detector consists of an array of optical detector cells. The resolution of frequency is directly related to the number of detector cells. High resolution capabilities are available through the use of TV pickup devices such as CCD optical detectors.

The Bragg cell receiver is capable of multiple signal handling, which produces individual outputs in the Fourier plane.

Summarizing, the Bragg cell receiver is an instantaneous processor capable of multiple signal handling, which can provide (as outputs) the frequency of the input frequencies, as well as their distribution. The frequency resolution of the system is directly related to the number of cells used in the detector array. The sensitivity of this type of system is in the vicinity of −80 dBm.

The limitation of this approach is the processing speed (which is limited by the photodetector response time and its inability to output signal characteristics for modulation analysis).

7

DESIGN EXAMPLES

Three examples are presented here, covering the most frequently encountered design problems. These are treated as they would be in practice. It is often necessary to modify a system design one or more times (as problems are encountered) in subsequent analysis. This is the advantage of performance analysis. Changes can be made on paper inexpensively before hardware is started. It is foolhardy to start hardware for a system, without prior performance analysis.

These examples are useful guides to the designer, who should follow the design sequence presented, after the configuration is selected. They assume that the initial system selection has been made. The reader is referred to the section on IF (4.3); which serves as background for these examples.

All of the computer programs are written in the BASIC language for the HP series 80 computers. Some minor modifications may be necessary for other machines.

7.1 EXAMPLE 1

Down Conversion Receiver

Specifications:	Required	Predicted
Frequency Range, MHz	30 to 80	30 to 80
IF Rejection, dB	> 60	> ≈ 80*
Image Rejection, dB	> 55	60
LO Radiation, dBm	< -80	-88
Detection Mode	FM	FM
Spurious Response, -dB	> 65	71
Intercept Point, dBm (3rd order, input)**	> 10	15.6
IF Bandwidth, kHz	30	30
Sensitivity, FM μv		
50Ω source open ckt.	14	7.1
$S+N/N \geqq 20$ dB		
Deviation ± 8kHz at 1 kHz		
Frequency Response, Audio	30 to 15 kHz	Same

*   The IF rejection is the sum of the preselector attenuation, 60 dB, plus the mixer's RF to IF isolation, which is 20 or more dB, for a total of 80 dB.

**   The truncation point for the intercept point is, in this case, the IF filter. This filter will stop the two tones if their spacing is greater than the filter bandwidth.

Selecting the Intermediate Frequency (IF)

For down conversion the following rules apply:

The IF must be out of band
The preselector must present its ultimate attenuation to the:

Intermediate frequency
Image frequency
Local oscillator frequency

To minimize the LO tuning range, high side injection is selected.

This is described by:

$$IF = F_{lo} - F_r$$

where

F_{lo} is the local oscillator frequency

and

F_r is the received frequency

Pictorially these conditions are shown in Fig. (7-1).

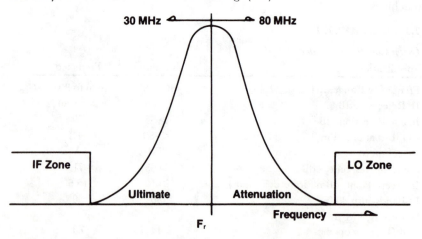

Fig. 7-1. Pictorial representation of the IF, LO and preselector relationships.

From Fig. (7-1) and the specifications, the IF must be less than 30 MHz (in the worst case, half the –60 dB bandwidth) of the preselector. Using two varactor-tuned critically coupled transformers in cascade for the preselector, the program of Table 5-4 is executed for the typical 4 to 12% bandwidth range. After several iterations it can be concluded that a 4% bandwidth is required for the preselector. This represents a loaded Q_l of $F_r/B_3 = 25$. The results are shown in Fig. (7-2). At 80 MHz, $F_{lo} \geq 95$ MHz, from which:

$$IF_{min} = 95 - 80 = 15 \text{ MHz}$$

Examining the low frequency end of $F_r = 30$ MHz, the highest IF which can be used is 25 MHz. From these limits an IF of 21.4 is selected and all of the rules are satisfied. (Note that the image frequency is above the local oscillator frequency by an amount equal to the IF.)

The Block Diagram

Having selected an IF and preselector design, the rough block diagram is drawn (as shown in Fig. (7-3)), using further inputs from the following considerations and calculations.

Sensitivity

The receiver must match the 50 Ω source impedance for best performance. The loaded input signal voltage to the receiver is $14\,\mu$volts $/\, 2 = 7\,\mu$volts. Converting this to dBm we have:

$$10 \log_{10} \frac{(7 \cdot 10^{-6})^2}{50} + 30 = -90 \text{ dBm}$$

Calculating the signal to noise ratio

$$S/N = 3 \left(\frac{\Delta F}{B_a} \right)^2 \frac{B_{if}}{2 B_a} \frac{C}{N}$$

$\Delta F = 8$ kHz peak
$B_a = 15$ kHz
$B_{if} = 40$ kHz(ENB)
$C = -90$ dBm
$N = kTB_{if} = -128$ dBm

$$S/N = (S+N)/N \cong 20 \ dB$$

$$S/N = 3 \left(\frac{8 \cdot 10^3}{15 \cdot 10^3} \right)^2 \left(\frac{40 \cdot 10^3}{2 \cdot 15 \cdot 10^3} \right) C/N$$

or

$$S/N = 1.138 \ C/N$$

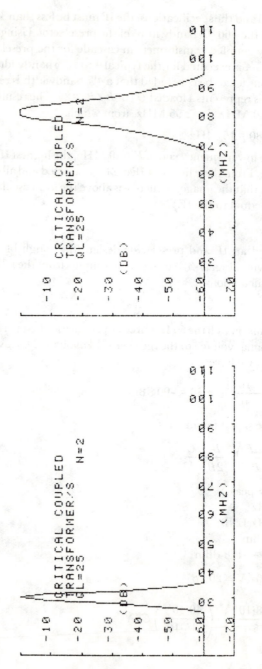

Fig. 7-2. Response of the preselector at 30 and 80 MHz showing the ultimate attenuation zones.

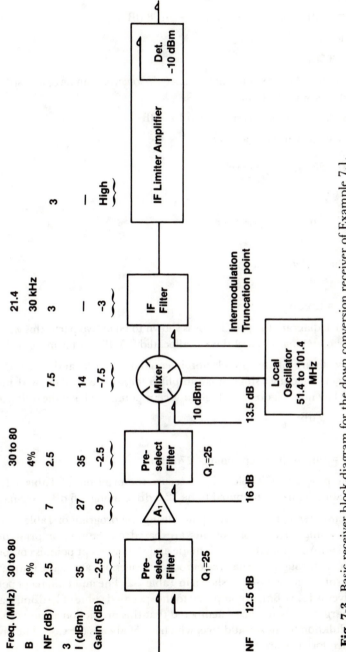

Fig. 7-3. Basic receiver block diagram for the down conversion receiver of Example 7.1.

In log from:

$S/N = 20$ dB $= 0.561 + (-90$ dBm$) - (-128$ dBm$) - NF$

Solving for the noise figure:

$NF = 18.561$ dB max

This represents the maximum noise figure the receiver can have and still meet the sensitivity specification.

The noise figure to the mixer input is 13.5 dB.

The preselector will have a loss of:

$$L \ (dB) = 20 \log_{10} \frac{Q_u}{Q_u - Q_l} \tag{7-1}$$

where

Q_u is the unloaded Q (assume 100)

and

Q_l is the loaded $Q = 25$

Then

L (dB) $= 2.5$ dB

In the block diagram the preselector is shown to be in two parts; this was done for better matching. Each of these is allocated 2.5 dB loss for margin.

The noise figure of the preselector is equal to its loss and is 2.5 dB per transformer. Selecting a single stage amplifier for the preamplifier with moderate gain, low noise figure, and high intercept point, we have the following:

Gain $= 9$ dB

NF_3 $= 7$ dB

I $=$ output intercept point $= 27$ dBm.

Using the program in Table 4-5, we secure the print-out of Table 7-1. The receiver noise figure is computed to be 12.5 dB, leaving a 6 dB margin.

The system intercept point is computed using the program of Table 4-6. The intercept points of the components are based on catalog or prior design information. An initial selection is made and the intercept point is computed. Adjustments through alternative component selections are made until a suitable fit results. The print-out is shown in Table 7-2. The input intercept point of the receiver is 15.6 dBm, as compared to the specified value of 10 dBm, leaving a 5 dB margin. It should be remembered that this prediction assumes that all intermodulation terms are additive, which is not always the case. The prediction is therefore pessimistic.

Table 7-1.

Computer Print-Out of the Noise Figure of the Down Converter Receiver

```
CASCADE NOISE FIGURE
 STARTS WITH LAST STAGE AND WORK
S UP TO THE INPUT
  NFT=10*LGT(F1+((F2-1)/G1));DB
  WHERE
 NFT=TOTAL NOISE FIGURE(DB)
 F1=PRECEDING STAGE NOISE FIGURE
(RATIO)
 F2=FOLLOWING STAGE NOISE FIGURE
(RATIO)
 G1=1ST STAGE GAIN
ALL PROGRAM ENTRIES ARE IN DB

***********************************
NF      G       CAS NF      STAGE

DB      DB      DB

 3      100             IF AMP
 3      -3      6       IF FLTR
 7.5    -7.5    13.5    MIXER
 2.5    -2.5    16      PRE SEL 1
 7      9       10      PRE AMP
 2.5    -2.5    12.5    PRE SEL 2
```

Table 7-2.

Computer Print-Out of the Third Order Input Intercept Point of the Down Converter Receiver

```
CASCADE INTERCEPT
 COMPUTES DEGRADATION OF THE IN
TERCEPT POINT DUE TO A PRECEDING
 STAGE

 THE INPUT INT PT IS THE OUTPUT
 INT PT-GAIN
CASCADE INTERCEPT POINT
THIRD ORDER
```

```
**********************************
IPT  G  CAS I   GT   IN IPT  STAGE
DBM  DB  DBM    DB    DBM

 35 -2.5                       PRE S

 27  9   26.9
                6.5   20.4    P AMP

 35 -2.5
         24      4     20     PRE S

 17 -7.5
         13.7
               -3.5   17.2    MXR
```

Spurious Response Predictions

These computations are best left to a computer. Utilizing the program of Table 4-7 and entering the appropriate parameters, a series of print-outs are obtained which show the spur frequency order and magnitude. To reduce the clutter, a spur search floor of 10 dB better than the specifications is used.

From the print-out of Table 7-3, we see that the worst spurious response is the $3 \times 1 = -71$ dB, resulting in a 6 dB margin and an image response of 60 dB (which results in a 5 dB margin).

The specifications are for a low performance receiver and are not generally satisfactory unless the receiver is operated in a weak signal environment. To improve the performance it is necessary to improve the preselector selectivity and ultimate attenuation by using traps.

The receiver noise floor is kTBF or

$-128 + 18.56 = -109.4$ dBm (specified or implied)

For the received signal to be below the noise floor, the image signal cannot exceed −49.4 dBm and the spurious signal input must be below −38.4 dBm. Stronger signals will cause these signals to be above the noise floor.

IF Rejection

From the print-out of Table 7-3, the IF rejection is better than the search floor of 70 dB. This response would have had an $M = 0$ and $N = 1$ identity.

IF Gain

To determine the gain required in the IF amplifier, the noise floor should be amplified up to the required detector level. For an FM application there are several detection forms to choose from, some of which require limiting. It remains to select one best suited to the task and then proceed with the calculation.

To illustrate, assume the detector requires an input of –10 dBm. The noise floor is –109.4 dBm. The difference is the required gain, which in this case is 99.4 dB. The pre-IF amplifier gain is –6.5 dB. Therefore the required IF gain is 106 dB. This is a rather high gain and it should be broken up into two frequencies, or the preamplifier gain should be increased by about 7 dB. An alternative solution is the use of a detector with a lower detection threshold. A good design rule is to at least break even in the pre-IF gain chain.

Table 7-3.

Spurious Prediction Print-Out for the Down Conversion Receiver of Example 7-1

```
SPUR SEARCH PROGRAM
RF INPUT -10DBM,LO 17 DBM
FROM
    FS=(FIF-M*FLO)/N
WHERE
    FS=SPUR FREQUENCY
    FIF=INTERMEDIATE FREQUENCY
    FLO=LOCAL OSC FREQ
    M&N ARE INTEGERS OF BOTH SIGN
S
COMPUTES UP TO 15TH ORDER

DEFINITIONS
    ORDER=ORD=ABS(M+N)
    FR=TUNED FREQUENCY
    FMIN=MIN LIMIT OF FR
    FMAX=MAX LIMIT OF FR

FMIN= 30                    FMAX= 60
IF= 21.4
OPTION= 1
SPUR FLOOR= 80 DB
TUNABLE FILTER USING 2
TRANSFORMERS WITH A LOADED Q OF
  25 AND ULT ATTN OF 60
```

```
**************************************
   TUNED FREQ= 30        FLO= 51.4
FSPUR        M    N   -DB   FLT   TOT
  338.4     -7    1    19    60    79
  235.6     -5    1    14    60    74
  132.8     -3    1    11    60    71
  30        -1    1     0     0     0
  72.8       1    1     0    60    60
  175.6      3    1    11    60    71
  278.4      5    1    14    60    74
  381.2      7    1    19    60    79
**************************************
   TUNED FREQ= 55        FLO= 76.4
FSPUR        M    N   -DB   FLT   TOT
  513.4     -7    1    19    60    79
  360.6     -5    1    14    60    74
  207.8     -3    1    11    60    71
  55        -1    1     0     0     0
  97.8       1    1     0    60    60
  250.6      3    1    11    60    71
  403.4      5    1    14    60    74
  556.2      7    1    19    60    79
**************************************
   TUNED FREQ= 80        FLO= 101.4
FSPUR        M    N   -DB   FLT   TOT
  688.4     -7    1    19    60    79
  485.6     -5    1    14    60    74
  282.8     -3    1    11    60    71
  80        -1    1     0     0     0
  122.8      1    1     0    60    60
  325.6      3    1    11    60    71
  528.4      5    1    14    60    74
  731.2      7    1    19    60    79
  118.9      9    8     0    54    54
```

Local Oscillator Radiation

The local oscillator signal path to the antenna is shown in Fig. (7-4). The local oscillator signal, whose magnitude is 10 dBm goes through the mixer L to R port isolation (–25 dB); the preselector (–30 dB); reverse isolation of the preamplifier (–13.5 dB); and the input preselector (–30 dB). This results in a LO level at the antenna of:

10 –25 –30 –13 –30 = –88 dBm

Where receivers are co-located with antennas in close proximity or in common, this may not be sufficient. In this case it would be necessary to improve the preselectivity. For this requirement the specifications are met with an 8 dB margin.

This example was presented to illustrate the method. The final design is left to the reader.

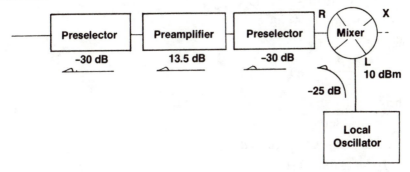

Fig. 7-4. The local radiation path for the receiver of Example 7.1.

7.2 **EXAMPLE 2**

Design a receiver covering the 50 to 1200 MHz frequency range, which meets the specifications of Table 7-4, using a hybrid up and down conversion scheme.

Table 7-4.

Specifications for EXAMPLE 2

Specifications:	Required	Predicted
Tuning Range, MHz	50 to 1200	50 to 1200
Sensitivity, dBm 30 AM at 1 kHz $(S+N)/N = 10$ dB	−90	−94.6
Bandwidth, kHz Shape Factor, 3 to 60 dB	25 3:1	25 3:1
LO Radiation, dBm	−80	−88
Image Rejection, dB	80	>80
IF Rejection, dB	80	80
Spurious Responses, dB	−75	−75 *
Intermodulation Distortion 3rd order 2 tones, −35 dBm spaced 1 MHz	−75	−80
AGC Cut On, dBm	−90	−100
AGC Range, dB	100	105

*F_{lo} + *2F*, Low end of band B

Sensitivity

$(S+N)/N = 10$ dB = 10 ratio

$S/N = 9$ ratio = 9.54 dB

and

$$S/N = \frac{P_c\, m^2}{2kTBF}$$

where

$P_c = -90$ dBm

$m = 0.3$, $m^2 = 0.09 = -10.46$ dB

$kT = -144$ dBm/kHz

$B = B_{if(ENB)} = 31$ kHz = 14.9 dB

F is the unknown

In dB notation:

$9.54 = -90 - 3 + (-10.46) - (-144) - 14.9 - NF$

$NF = 16.1$ dB

This is the maximum system noise figure. The maximum noise floor level is kTBF or:

$-144 + 14.9 + 16.1 = -113$ dBm

System Configuration

The tuning range of 50 to 1200 MHz is split into two parts, one of which will be down converted and the other up converted.

This is defined as follows:

Band *A*

$F_{min(A)} = 50$ MHz
$F_{max(A)} = 50 + (1200 - 50)/2 = 625$ MHz

and

Band *B*

$F_{min(B)} = 625$ MHz
$F_{max(B)} = 1200$ MHz

This spectral division is shown in Fig. (7-5).

Band *A* will be up converted into an IF in band *B*, and band *B* will be down converted into an IF in band *A*. By symmetric spacing of the IFs the first LO tuning range is reduced by 1/2. This spacing must be made equal to the width of the bands, i.e., 575 MHz. Then

Band A IF $= 625 + (625 - 50)/2 = 912.5$ MHz

and

Band B IF $= 625 - (625 - 50)/2 = 337.5$ MHz

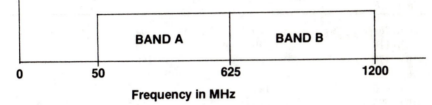

Fig. 7-5. RF band plan for the frequency range of 50 to 1200 MHz.

With this selection, the frequency relationships are tabulated in Table 7-5.

Table 7-5.
Receiver Frequency Relationships (MHz)

Band	Receive Frequency	IF	1st LO	Mode
A	50	912.5	962.5	IF $= F_{lo} - F_r$
A	625	912.5	1537.5	IF $= F_{lo} - F_r$
B	625	337.5	962.5	IF $= F_{lo} - F_r$
B	1200	337.5	1537.5	IF $= F_{lo} - F_r$

Note that the same LO is used in both bands.

Preselector

Because band A is up converted, it is acceptable to use a fixed tuned filter bank. The number of filters required is:

$$50 k^n = 625$$
$$k = \sqrt[n]{12.5}$$

An n value of 6 gives a $k = 1.52$, which is close to the 1.5 rule for four section filters. Thus, six filters will be used. These are listed in Table 7-6.

A tentative filter selection is made at this point to be a 0.1 dB ripple Chebyshev filter with an ultimate attenuation of 60 dB.

Band B (although down converted) is a border line case and fixed tuned filtering will be used. This results because of the relatively high band B IF. For the preselector to be effective in suppressing LO radiation, the filter's ultimate attenuation must be presented to the LO. With a fixed k value, preselector bandwidth increases with frequency. The $k = 1.5$ rule does not apply here. The worst case would occur at 1200 MHz.

Table 7-6.

Band *A* Preselector Specifications (MHz)

(F to 1.52F)	Width	F_o	% B
50 to 76	26	63	41
76 to 116	40	96	41
116 to 176	60	146	41
176 to 269	93	225.5	41
269 to 410	141	339.5	41
410 to 625	215	517.5	41

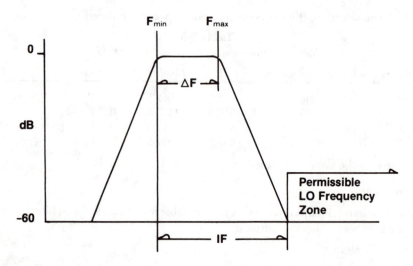

Fig. 7-6. Desired preselector frequency relationships for band *B* (625 to 1200 MHz).

Executing the program of Table 5-6, and iterating until the LO is related to the bandwidth, as shown in Fig. (7-6), it is found that the bands can be defined as follows:

Table 7-7.

Band B Preselector Specifications (MHz)

Frequency (MHz)					Ripple (dB)	% B
Min	Max	ΔF	N	F_o		
1060	1200	140	5	1130	0.1	12.4
915	1060	145	5	987.5	0.1	14.6
770	915	145	5	842.5	0.1	17.2
625	770	145	5	697.5	0.1	20.0

Preselector filter definitions for the down conversion of 625 to 1200 MHz frequencies are given in Table 7-7. ΔF is the bandwidth of the filter in MHz, N is the number of sections, F_o is the center frequency, and $\%B$ is the bandwidth in percent.

The system block diagram may now be drawn and the related information computed and added. This is shown in Fig. (7-7). The preselector filter bank shown in the system block diagram of Fig. (7-7) is expanded in Fig. (7-8). It consists of a 10 PST pin diode switch feeding 10 filters whose outputs are selected by a second 10 PST switch. All filters are 0.1 dB ripple Chebyshev types using four or five sections as shown. The worst case loss and noise figure is 3.5 dB and a third order intercept point is 35.5 dB.

Preamplifier

A single preamplification stage is selected with a gain of 12 dB, a noise figure of 2 dB and an output intercept point (third order) of 20 dBm. This is an initial choice and the final selection will be contingent upon the overall system performance. Experience will serve as a guide in this initial selection. Where a high intercept point is called for, a low gain power type of stage is required. These generally have a poorer noise figure. In this case the selection made should be adequate.

This amplifier is placed between the preselector and the lowpass filter. The amplifier serves as a termination to both filters and results in a good VSWR.

The Mixers

Both mixers are double balanced for good isolation and spurious performance. In all but low performance receivers, the double balanced mixer should be used. For best spurious and intermodulation performance this mixer should be of the termination insensitive type, operating with high LO drive (17 to 23 dBm typical). For this design a drive level of 17 dBm is selected for both mixers, which are diode double balanced, termination insensitive types.

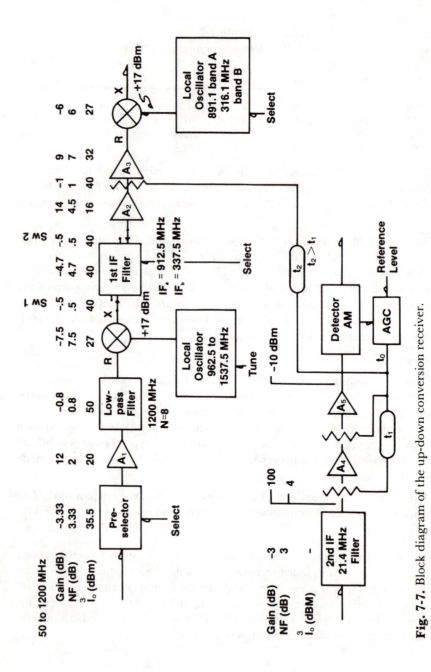

Fig. 7-7. Block diagram of the up-down conversion receiver.

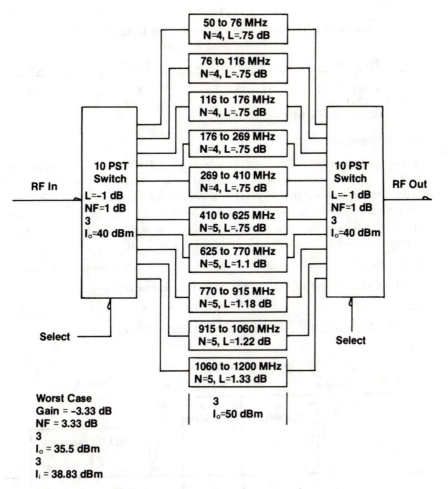

Fig. 7-8. Preselector configuration and performance data.

The Second Local Oscillator

The second local oscillator is selected for low side injection because lower frequencies are less noisy. The second LO frequencies are related to the receive and first intermediate frequencies, as shown in Table 7-8.

For the best frequency accuracy, these frequencies should be phase locked to the first LO reference standard.

Table 7-8.

Frequency Relationships Selected for the Hybrid Up-Down
Conversion Receiver

Receive Frequency	1st IF	2nd IF	2nd LO
50 to 625	912.5	21.4	891.1
625 to 1200	337.5	21.4	316.1

All values are shown in MHz.

First IF Amplifier

The first IF amplifier is driven by the first mixer. A SPDT switch drives either of
two IF filters, whose outputs are selected by a second SPDT switch. These
filters are made as narrow as practical, to improve intermodulation performance and local oscillator feed through. The critical requirements result from
the two local oscillator frequencies, which must be in the first IF filter ultimate
attenuation zone. Pictorially this is shown in Fig. (7-9).

From this, the critical first IF filter requirements are determined as follows:

316.1 MHz = first IF band *B* filter ultimate attenuation

and

891.1 MHz and 962.5 = first IF band *A* filter ultimate attenuation.

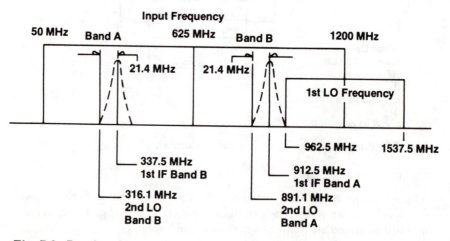

Fig. 7-9. Receiver frequency relationships for RF, IF, and LO frequencies.

For the best system noise figure, this filter should have low loss. This is achieved by minimizing the number of sections. By either referring to a catalog or the program in Table 5-6, the filter can be defined. For a four section 0.05 dB Chebyshev filter the 70 dB bandwidth (of the filter) is about eight times the 3 dB bandwidth. Then, four bandwiths must equal to < 21.4 MHz or $B \approx 5$ MHz. This represents a 1.48% bandwidth. Such a narrow bandwidth dictates the use of a cavity-type filter. There the insertion loss varies with size. An approximate value of loss constant of 1.4 to 1.2 will be used.

Then

Insertion loss (337.5 MHz filter) =

$$\frac{\left(\text{loss constant}\right) \left(\text{No. of sections} + \frac{1}{2}\right)}{\% \ 3 \ \text{dB Bandwidth}} + 0.2$$

$$= \frac{(1.4 \ (4 + 0.5)}{1.48} + 0.2 = 4.7 \ \text{dB}$$

Similarly for band A (the 962.5 MHz filter), the 3 dB bandwidth is ≈ 5 MHz or 0.5% and the insertion loss is:

$$L = \frac{0.35 \ (4 + 0.5)}{0.5} + 0.2 = 3.35 \ \text{dB}$$

The IF filters are followed by two amplifiers in cascade separated by an AGC attenuator. The gain of these amplifiers should be sufficient to cancel any pre-second IF amplifier loss.

Computing the required gain between the detector and the antenna we have:

Detector Signal = Total Gain + Noise Floor (dB notation)

–10 dBm = Total Gain + kTBF

From which

Total Gain = 113 dBm – 10 dBm = 103 dB

For good intermodulation performance it is necessary to minimize the gain ahead of the truncating IF filter, which in this case is the 21.4 MHz second IF filter. Therefore the second IF gain should be maximized (but it should be less than the practical limit of 100 dB). In this example the pre-second IF gain must be 3 dB. Summing the losses and gains to the second IF filter we have:

Table 7-9.

Tabulation of Gains and Losses of the Pre-second IF Amplifier to Compute the
Gains of *A1, A2,* and *A3*

Stage	Loss (dB)	Gain (dB)
Preselector	3.33	
Preamplifier		A1 (12)
Lowpass Filter	0.8	
1st Mixer	7.5	
1st IF switch	0.5	
1st IF Filter	4.7	
1st IF Switch	0.5	
1st IF Amplifier		A2 (14)
AGC Block	1.0	
1st IF Amplifier		A3 (9)
2nd Mixer	6.0	
2nd IF Filter	3.0	
	27.33	3 + 27.33, –0 + 10
		(Required gain)

Amplifiers *A1, A2,* and *A3* must total a gain of 30.33 – 0 + 10 dB. Amplifier *A1* is
12 dB and amplifiers *A2* and *A3* are budgeted at 14 dB and 9 dB respectively,
giving a 5 dB margin.

The Second IF Amplifier

This amplifier is driven through the second mixer and the IF crystal filter. The
AGC attenuators are of the PIN diode variety. Each of these provide up to >35
dB of attenuation for a total of > 105 dB (including the one located in the first
IF chain). They are distributed in the gain chain, such that at no time does any
stage go into compression on modulation peaks with the maximum signal to be
encountered. Because the gain was computed to noise threshold, the AGC
system will begin to function with any noise, as required. The second IF
amplifier contains the bulk of the IF gain of 100 dB. The remaining 3 dB plus
margin is secured in pre-second IF gain.

Image Rejection

There are two image cases to consider, both of which are defined by;

$$IF = | F_{lo} - F_r |$$

For the 50 to 625 MHz receive band:

$$912.5 = | (962.5 \text{ to } 1537.5) - F_r |$$

There are two solutions:

Desired F_r = 50 to 625 MHz and

Image F_r = 1875 to 2450 MHz

The undesired image frequencies are out of band and attenuated by the preselector and lowpass filter.

For the 625 to 1200 MHz receive band:

$337.5 = |\ (962.5\ \text{to}\ 1537.5) - F_r\ |$

The solutions are:

Desired F_r = 625 to 1200 MHz

Image F_r = 1300 to 1865 MHz

This undesired image response is also out of band and attenuated by the preselector and lowpass filters. The preselector can supply 60 dB of image attenuation which by itself is not enough. Therefore the lowpass filter must be added to the preselector.

For the 50 to 625 MHz range the lowpass filter will provide its ultimate attenuation of > 70 dB. For the 625 to 1200 MHz range, the lowpass filter is less effective toward the low edge of this band. However, only 20 dB of additional attenuation is needed at 1300 MHz. A sharp cutoff 1200 MHz lowpass filter will have to be used to secure this value. An eight section filter will provide this performance with only 0.8 dB thus of insertion loss. Therefore image rejection will range from 130 dB to > 80 dB, thus meeting the requirements.

IF Rejection

The first IF is always out of the preselector band. Therefore the preselector will provide its ultimate attenuation of 60 dB to this frequency. Additionally, since the response at the IF is not a converted response, the first mixer will provide an additonal attenuation of 20 to 30 dB; resulting from its input/output (R to X) port leakage. The total will be in excess of 80 dB. The configuration will meet the requirements of > 80 dB.

LO Radiation

The LO radiation path is shown in Fig. (7-10). The LO, whose magnitude is 17 dBm, goes through the LO to R_f port leakage of 20 to 30 dB. The following lowpass filter, whose cutoff is 1200 MHz, is ineffective in suppressing LO frequencies between 962.5 and 1200 MHz. The signal is then attenuated by the preamplifier reverse isolation of 25 dB and fed to the preselector, which by design provides 60 dB of additional loss. The net result is a LO level of –88 dBm at the antenna (which meets the requirements with an 8 dB margin).

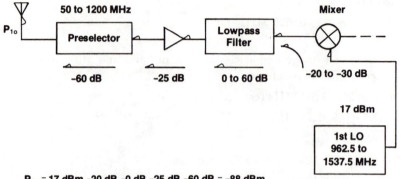

P_{lo} = 17 dBm –20 dB –0 dB –25 dB –60 dB = –88 dBm

Fig. 7-10. The local oscillator leakage path with a worst case P_{lo} level of –88 dBm.

Noise Figure

The computer program of Table 4-5 was utilized with the inputs taken from the block diagram of Fig. (7-7). The result is shown in the print-out of Table 7-10. The design is predicted to have an overall noise figure of 11.5 dB, compared to the required value computed to be 16.1 dB. The sensitivity of the design is therefore –90 – (16.1 – 11.5) = –94.6 dBm.

Intermodulation Distortion

The specification calls for the intermodulation distortion products to be $>$ –75 dB below the two tone input level of –35 dBm, where the tones are separated by 1 MHz. Because of 1 MHz spacing, the two-tone truncation point is the second IF filter.

The required input intercept point is from Figures (4-22) and (4-23), or computation, to be 2.5 dBm. Exercising program Table 4-6 and entering the values from Fig. (7-7), the computer print-out of Table 7-11 indicates a third order system intercept point of 5 dBm, which meets the requirements with a 2.5 dB margin.

Spurious Response

The computer program of Table 4-7 is executed using inputs from Fig. (7-7) and (7-8). For a complete prediction a minimum of 20 runs should be made. (There are 10 preselector filters which should be examined, as a minimum, at each band edge.) By using a floor 5 dB below the requirements, the print-out is less cluttered.

Table 7-10.
Computer Print-Out of the System Noise Figure for the
Hybrid Conversion Receiver

```
CASCADE NOISE FIGURE
 STARTS WITH LAST STAGE AND WORK
S UP TO THE INPUT
  NFT=10*LGT(F1+((F2-1)/G1)),DB
  WHERE
 NFT=TOTAL NOISE FIGURE(DB)
 F1=PRECEDING STAGE NOISE FIGURE
(RATIO)
 F2=FOLLOWING STAGE NOISE FIGURE
(RATIO)
 G1=1ST STAGE GAIN
ALL PROGRAM ENTRIES ARE IN DB


*********************************
NF      G      CAS NF      STAGE

DB      DB      DB

 4      100              2ND IF AMP
 3      -3      7        2ND IF FLTR
 6      -6      13       2ND MXR
 7      9       8 7      AMP A3
 1      -1      9.7      AGC ATTN
 4.5    14      5        AMP A2
 .5     -.5     5.5      IF SW
 4.7    -4.7    10.2     IF FLTR
 .5     -.5     10.7     IF SW
 7.5    -7.5    18 2     1ST MXR
 .8     -.8     19       LOW PASS
 2      12      8.1      PRE AMP
 3.33   -3.33   11 5     PRE SELECT
```

Table 7-11.
Computer Print-Out of the Cascade Third Order Input Intercept Point
for the Hybrid Conversion Receiver

```
CASCADE INTERCEPT
 COMPUTES DEGRADATION OF THE IN
TERCEPT POINT DUE TO A PRECEDING
 STAGE
```

THE INPUT INT PT IS THE OUTPUT
INT PT-GAIN
CASCADE INTERCEPT POINT
THIRD ORDER
****

IPT DBM	G DB	CAS I DBM	GT DB	IN IPT DBM	STAGE
35.5			-3.33		PRE S
20	12	20	8.67	11.33	AMP 1
50	-.8	19.2	7.87	11.33	LOW P
27	-7.5	11.6	.37	11.23	MIX 1
40	-.5	11.1	-.13	11.23	SW
40	-4.7	6.4	-4.83	11.23	IFFLT
40	-.5	5.9	-5.33	11.23	SW
16	14	14.5	8.67	5.83	AMP 2
40	-1	13.5	7.67	5.83	AGC
32	9	22	16.67	5.33	AMP 3
27	-6	15.7	10.67	5.03	MIX 2

For this example only four filters are examined (eight runs) with a floor of −80 dB. These are shown in Table 7-12. There appears to be a problem in the 625 to 770 MHz band on the image and the 1 x 2 spur. The 1 x 2 spur is right at specifications and is in band as a consequence of the system, with no solution for improvement. This should be flagged and verified on the bench. The image is actually in spec (since it was pointed out that a fast fall filter will be used which is not in the program). Actually the fast fall filter will add ~ 30 dB loss rather than the seven shown on the run, totaling ~ 90 dB down.

Table 7-12.

Computer Print-Out of the Spurious Performance of the Hybrid Conversion Receiver

```
SPUR SEARCH PROGRAM
RF INPUT -10DBM,LO 17 DBM
FROM
    FS=(FIF-M*FLO)/N
WHERE
    FS=SPUR FREQUENCY
    FIF=INTERMEDIATE FREQUENCY
    FLO=LOCAL OSC FREQ
    M&N ARE INTEGERS OF BOTH SIGN
S
COMPUTES UP TO 15TH ORDER

DEFINITIONS
    ORDER=ORD=ABS(M+N)
    FR=TUNED FREQUENCY
    FMIN=MIN LIMIT OF FR
    FMAX=MAX LIMIT OF FR

FMIN= 50               FMAX= 76
IF= 912.5
OPTION= 1
SPUR FLOOR= 80 DB
LOW PASS FILTER CUT OFF= 1200
NUMBER OF ELEMENTS = 8
RIPPLE= .1 DB
ULTIMATE ATTN= 70 DB
FIXED TUNED CHEBYSHEV FILTER FMI
N= 50 FMAX= 76 N= 4 RIPPLE= .1
ULT ATN= 60
*********************************
   TUNED FREQ= 50       FLO= 962.5
FSPUR     M    N    -DB   FLT    TOT
 50       -1    1    0     0      0
```

```
*******************************
   TUNED FREQ= 76        FLO= 988.5
FSPUR        M    N   -DB   FLT   TOT
  76        -1    1    0     0     0




FMIN= 410                      FMAX= 625
IF= 912.5
OPTION= 1
SPUR FLOOR= 80 DB
LOW PASS FILTER CUT OFF= 1200
NUMBER OF ELEMENTS = 8
RIPPLE= .1 DB
ULTIMATE ATTN= 70 DB
FIXED TUNED CHEBYSHEV FILTER FMI
N= 410 FMAX= 625 N= 5 RIPPLE=
.1 ULT ATN= 60
*******************************
   TUNED FREQ= 410       FLO=
1322.5
FSPUR        M    N   -DB   FLT   TOT
 410        -1    1    0     0     0
*******************************
   TUNED FREQ= 625       FLO=
1537.5
FSPUR        M    N   -DB   FLT   TOT
 625        -1    1    0     0     0


   SPUR SEARCH PROGRAM
   RF INPUT -10DBM,LO 17 DBM
   FROM
      FS=(FIF-M*FLO)/N
   WHERE
      FS=SPUR FREQUENCY
      FIF=INTERMEDIATE FREQUENCY
      FLO=LOCAL OSC FREQ
      M&N ARE INTEGERS OF BOTH SIGN
   S
   COMPUTES UP TO 15TH ORDER

   DEFINITIONS
      ORDER=ORD=ABS(M+N)
      FR=TUNED FREQUENCY
      FMIN=MIN LIMIT OF FR
      FMAX=MAX LIMIT OF FR
```

```
FMIN= 625               FMAX= 770
IF= 337.5
OPTION= 1
SPUR FLOOR= 80 DB
LOW PASS FILTER CUT OFF= 1200
NUMBER OF ELEMENTS = 8
RIPPLE= .1 DB
ULTIMATE ATTN= 70 DB
FIXED TUNED CHEBYSHEV FILTER FMI
N= 625 FMAX= 770 N= 5 RIPPLE=
.1 ULT ATN= 60
************************************
  TUNED FREQ= 625      FLO= 962.5
FSPUR       M    N   -DB   FLT   TOT
 625       -1    1    0     0     0
 1300       1    1    0    60    67
 650        1    2   75     0    75
************************************
  TUNED FREQ= 770      FLO=
1107.5
FSPUR       M    N   -DB   FLT   TOT
 770       -1    1    0     0     0
 722.5      1    2   75     0    75



FMIN= 1060              FMAX= 1200
IF= 337.5
OPTION= 1
SPUR FLOOR= 80 DB
LOW PASS FILTER CUT OFF= 1200
NUMBER OF ELEMENTS = 8
RIPPLE= .1 DB
ULTIMATE ATTN= 70 DB
FIXED TUNED CHEBYSHEV FILTER FMI
N= 1060 FMAX= 1200 N= 5 RIPPLE=
.1 ULT ATN= 60
************************************
  TUNED FREQ= 1060     FLO=
1397.5
FSPUR       M    N   -DB   FLT   TOT
 1060      -1    1    0     0     0
************************************
  TUNED FREQ= 1200     FLO=
1537.5
FSPUR       M    N   -DB   FLT   TOT
 1200      -1    1    0     0     0
```

7.3 EXAMPLE 3

Design a receiver using an up conversion system covering the frequency range of 30 to 250 MHz. Present designs using standard up conversion and the Wadley system. Since the details of computing performance are presented in Example 2, limit this effort for a receiver with spurious performance of –80 dB with inputs of –10 dBm to only the preselector and IF choice.

Preselector

Either a fixed or varactor tuned filter bank is acceptable. Rough calculations indicate that six fixed tuned filters are required for the fixed filter band, while only three are necessary for the varacter tuned filter bank. Also the varacter tuned filter provides superior spurious performance. Choosing the latter, the tuning ranges are:

$$30k^n = 250$$
$$k = \sqrt[n]{250 / 30}$$

where

$k \approx\, < 2$, $n = 3$ and $k = 2.027$

The filters are defined as follows:

Frequency (MHz)	Number of transformers (N)	Bandwidth %
30 to 61	2	5
61 to 123	2	5
123 to 250	2	5

IF Selection

The first IF must be > 3 times the higest tuned frequency to prevent $3F_r = IF_1$ from being < 80 dB down. Then $IF_1 > 3.250 = > 750$ MHz. A tentative value is 790 MHz which allows a 40 MHz guard band between these frequencies within which the first IF filter will be at its ultimate attenuation value at $3F_{rmax}$.

Second IF Selection

Standard Up Converter

The second IF is selected to be the popular frequency of 21.4 MHz where filters are readily available and the frequency is low enough to allow a high gain amplifier to be used. Any IF less than that of the first IF could have been used, but it is good practice to get to a workable frequency with as few conversions as possible. Other possibilities are 30, 70, 120, *et cetera.* A 10.7 MHz value would demand a very narrow first IF filter which would have higher loss and poorer stability.

The First Local Oscillator

The first LO frequency range is

$$F_{lo_1} = |\ IF_1 \pm F_r\ |$$

where

IF$_1$ = 790 MHz
F_r = 30 to 250 MHz

There are two choices:

$$F_{lo_1} = 790 + (30 \text{ to } 250) = 820 \text{ to } 1040 \text{ MHz}$$
or
$$F_{lo_1} = 790 - (30 \text{ to } 250) = 760 \text{ to } 540 \text{ MHz}$$

Of these the lower frequency range is selected.

The Second Local Oscillator

The second LO could be placed on either side of the first IF without serious conflict. However, for maximum isolation this frequency is placed on the high side of the first IF. (Low side injection would have had a worst case separation of only 8.6 MHz from the maximum frequency of the first LO.)

Then

$$F_{lo_2} = IF_1 + IF_2$$
$$= 790 + 21.4 = 811.4 \text{ MHz}$$

The First IF Filter

The principal constraint is that the first IF filter be non-responsive (present its ultimate attenuation) to frequencies removed from its center frequencies, by an amount equal to the second IF. Fig. (7-11) illustrates this case.

From Fig. (7-11) the ultimate attenuation bandwidth must be less than $2 \cdot IF_2 =$ 42.8 MHz. A four section cavity filter has a –70 dB bandwidth of $\sim \pm 4$(–3dB bandwidths). Then the 3 dB bandwidth is:

$$B_3 = 42.8/8 = 5.35 \text{ MHz max.} = 0.68\%$$

A 0.5% bandwidth is therefore selected, with a ripple of .05 dB.

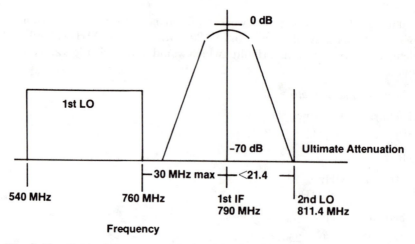

Fig. 7-11. Critical frequency relationships for Example 3.

Basic Block Diagram of the Up Converter Receiver

At this point the basic block diagram may be drawn; it is shown in Fig. (7-12).

The Wadley Design

The preselector and first IF are unchanged from that of Fig. (7-12). Defined are:

> Preselector, same as Fig. (7-12)
> First IF = 790 MHz
> First IF Filter = 790 MHz, $N = 4$, $B = 0.5\%$, 0.05 dB ripple cavity

The basic block diagram of Fig. (7-13) is drawn and completed as the following are considered.

The Second Intermediate Frequency

It is good practice to keep the variable frequency oscillator (VFO) in the ultimate attenuation zone of the preselector. The preselector is widest at 250 MHz. Then with a 5% bandwidth and an ultimate attenuation bandwidth of ±4 (–3 dB bandwidths), the closest this oscillator can be to the receive frequency is 50 MHz. Therefore, the second IF must be > 50 MHz. Since 70 MHz is a frequency where filters are available, this value is selected.

Wadley Oscillator Frequency (Second Local Oscillator)

The Wadley oscillator frequency is:

$$F_{lo_2} = \text{first IF} + \text{second IF} = 790 + 70 = 860 \text{ MHz}$$

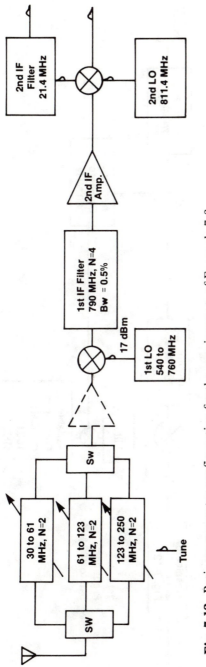

Fig. 7-12. Basic up converter configuration for the requirements of Example 7-3.

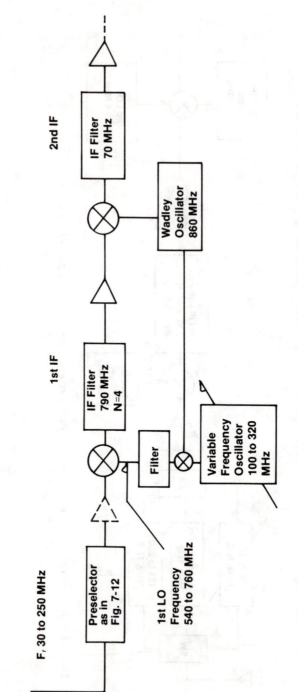

Fig. 7-13. Basic Wadley up conversion system for the reception of 30 to 250 MHz signals in Example 7-3.

Note that the difference frequency was not used because it would fall into the first LO frequency range of 540 to 760 MHz. This avoids potential beats and whistles.

The Variable Frequency Oscillator (VFO)

The VFO must satisfy the relationship:

first LO frequency = Wadley frequency – VFO frequency

then

(540 to 760) = 860 – VFO frequency

and

VFO frequency = 100 to 320 MHz

Note that the + sign was not used because the result would have been a VFO frequency in the GHz range, which becomes more difficult to realize, particularly if phase locking is used.

The First Local Oscillator

The VFO and Wadley oscillator frequencies are combined in a mixer and the frequency difference is selected by an appropriate filter, producing the first LO frequencies of 540 to 760 MHz. Note that where it is desirable to reduce the VFO tuning range by 50%, the second IF would have been

$$\frac{|F_{r_{min}} - F_{r_{max}}|}{4} = \frac{|30 - 250|}{4} = 55 \text{ MHz}$$

Then by using both the sum and difference modes in the first LO mixer, the VFO range would have been as follows:

The input RF band would be split into two equal parts.

Part 1

$$F_{r_{min}} \text{ to } \left(F_{r_{min}} + \frac{|F_{r_{min}} - F_{r_{max}}|}{2} \right)$$

and

Part 2

$$\left(F_{r_{min}} + \frac{|F_{r_{min}} - F_{r_{max}}|}{2} \right) \text{ to } F_{r_{max}}$$

Or

30 to 140 and 140 to 250 MHz

The frequency relationships would be as shown in Table 7-13.

Table 7-13.

Frequency Relationships for a Wadley Configuration (MHz)

Receiver Frequency	VFO Frequency	1st LO Frequency	Wadley Mixer Mode	Wadley Oscillator Frequency	2nd IF
30	85	760	–	845	55
140	195	650	–	845	55
>140	85	930	+	845	55
250	195	1040	+	845	55

From this it can be seen that the VFO range has been reduced by 50%. The penalty paid is usually a need for three conversions or a non-standard IF. Also two sideband filters must be switched in the LO chain.

The Wadley technique results in the generation of a first LO which is contaminated by spurious frequencies, because of the mixing process involved. This results in a receiver which has poorer spurious performance than obtained with standard up conversion.

APPENDIX

(a)

Digital Data Rate

Two terms are used to describe the data transmission speed of a digital system. These are bits per second and baud. A bit represents the smallest piece of information transmitted. The baud is one signal element per unit time, usually taken as one second.

Data to be transmitted must be encoded to expand the system's vocabulary and provide some means of identifying start and stop. Thus, as an example, to transmit n useful bits, $n + k$ bits would have to be used. The number of useful message bits per unit time is the system speed in bits per unit time. The total bits transmitted are baud rate or bauds per unit time.

(b)

Adding Numbers (dB) Notation

To add numbers in dB notation it is first necessary to take the anti-logarithm of the numbers, add them and then convert that sum to dB form. While not difficult, this can be time consuming, especially if many numbers are involved. Table b-1 and the curve of of Fig. (b-1) may be used to perform the task of adding in dB notation.

To use the table or curve, enter at the dB difference and read the add to larger value, then add it to the larger dB number.

Example b-1:

Add two powers of 20 and 26 dBm. The difference is 6 dB. From the table or chart find .96 dB, and add it to the larger. The sum is then 26 + .96 or 26.96 dBm.

Table b-1.
Addition of Units in dB Notation

DIFFERENCE (DB)	ADD TO LARGER (DBM)
0	3.01
.5	2.77
1	2.54
1.5	2.32
2	2.12
2.5	1.94
3	1.76
3.5	1.6
4	1.46
4.5	1.32
5	1.19
5.5	1.08
6	.97
6.5	.88
7	.79
7.5	.71
8	.64
8.5	.57
9	.51
9.5	.46
10	.41
10.5	.37
11	.33
11.5	.3
12	.27
12.5	.24
13	.21
13.5	.19
14	.17
14.5	.15
15	.14
15.5	.12
16	.11
16.5	.1
17	.09
17.5	.08
18	.07
18.5	.06
19	.05
19.5	.05
20	.04

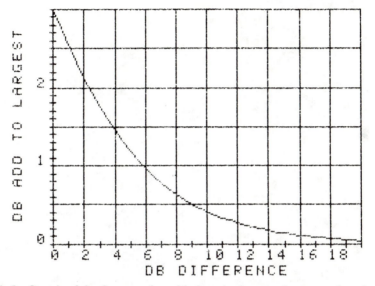

Fig b-1. Graph of the factor to be added to the larger of two numbers in dB notation when addition is performed.